COMPUTER AIDED DECISION SUPPORT IN TELECOMMUNICATIONS

BT Telecommunications Series

The BT Telecommunications Series covers the broad spectrum of telecommunications technology. Volumes are the result of research and development carried out, or funded by, BT, and represent the latest advances in the field.

The series includes volumes on underlying technologies as well as telecommunications. These books will be essential reading for those in research and development in telecommunications, in electronics and in computer science.

1. *Neural Networks for Vision, Speech and Natural Language*
   Edited by R Linggard, D J Myers and C Nightingale

2. *Audiovisual Telecommunications*
   Edited by N D Kenyon and C Nightingale

3. *Digital Signal Processing in Telecommunications*
   Edited by F A Westall and S F A Ip

4. *Telecommunications Local Networks*
   Edited by W K Ritchie and J R Stern

5. *Optical Network Technology*
   Edited by D W Smith

6. *Object Oriented Techniques in Telecommunications*
   Edited by E L Cusack and E S Cordingley

7. *Modelling Future Telecommunications Systems*
   Edited by P Cochrane and D J T Heatley

8. *Computer Aided Decision Support in Telecommunications*
   Edited by P G Flavin and K A E Totton

COMPUTER AIDED DECISION SUPPORT IN TELECOMMUNICATIONS

Edited by

P.G. Flavin
BT Laboratories
Martlesham Heath
UK

and

K.A.E. Totton
BT Laboratories
Martlesham Heath
UK

CHAPMAN & HALL
London · Weinheim · New York · Tokyo · Melbourne · Madras

Published by Chapman & Hall, 2–6 Boundary Row, London SE1 8HN, UK

Chapman & Hall, 2–6 Boundary Row, London SE1 8HN, UK

Chapman & Hall GmbH, Pappelallee 3, 69469 Weinheim, Germany

Chapman & Hall USA, 115 Fifth Avenue, New York, NY 10003, USA

Chapman & Hall Japan, ITP-Japan, Kyowa Building, 3F, 2-2-1 Hirakawacho, Chiyoda-ku, Tokyo 102, Japan

Chapman & Hall Australia, 102 Dodds Street, South Melbourne, Victoria 3205, Australia

Chapman & Hall India, R. Seshadri, 32 Second Main Road, CIT East, Madras 600 035, India

First edition 1996

© 1996 British Telecommunications plc
Softcover reprint of the hardcover 1st edition 1996
ISBN-13: 978-94-010-6524-5   e-ISBN-13: 978-94-009-0081-3
DOI: 10.1007/978-94-009-0081-3

A catalogue record for this book is available from the British Library

∞ Printed on permanent acid-free text paper, manufactured in accordance with ANSI/NISO Z39.48-1992 and ANSI/NISO Z39.48-1984 (Permanence of Paper).

Contents

Contributors

P R Benyon   Intelligent Systems, BT Laboratories

P M Bull   Intelligent Systems, BT Laboratories

S L Corley   Advanced Specification, BT Laboratories

I B Crabtree   Intelligent Systems, BT Laboratories

N J Davies   Intelligent Systems, BT Laboratories

R G Davison   Intelligent Systems, BT Laboratories

P G Flavin   Intelligent Systems, BT Laboratories

M T Last   ATM Futures, BT Laboratories

P R Limb   Intelligent Systems, BT Laboratories

G J Meggs   Intelligent Systems, BT Laboratories

D Munaf   Formerly Intelligent Systems, BT Laboratories

I Napier   Formerly Financial Analyst, BT London

P D O'Brien   Intelligent Systems, BT Laboratories

S O'Donnell   Formerly Intelligent Systems, BT Laboratories

M C Revett   Intelligent Systems, BT Laboratories

P M Sanders   Service Creation, BT Laboratories

R T Scarfe   Intelligent Systems, BT Laboratories

D L Scott   Intelligent Systems, BT Laboratories

R J Shortland   Customer Handling Systems, BT Laboratories

G D Tattersall   Field Support, BT Laboratories

K A E Totton   Intelligent Systems, BT Laboratories

P C Utton   Decision and Performance, BT Laboratories

Preface

'Never tell people how to do things. Tell them what to do and they will surprise you with their ingenuity.'

General George Patton, *War as I Knew It* (1947)

This quotation from Patton relating his wartime experience has many parallels in today's business environment. It also serves to introduce the subject of Computer Aided Decision Support. It illustrates succinctly the quintessence of decision support systems — that they are designed to assist people in establishing the best course of action in a given situation but not to automate or prescriptively tell them how to achieve a goal.

Half a century on from Patton's time, we have developed computing technology that brings immense benefits to our business lives. We are surrounded by colossal amounts of information and have the ability to process it at enormous speed. Compared with the limited data and rudimentary models that were available to Patton, who had to make decisions affecting thousands of lives, one could be forgiven for thinking that human decision making should now be more or less infallible. Clearly this is not actually the case. We now live in a remarkably complex world where there is a high degree of inter-dependence between seemingly unrelated issues. Take for example the environmental impact of most human activity — a world where, through the application of modern communications technology, the impact of a single decision can be felt around the world instantly, a world where there is so much information available that often no single person can know enough to take a rational decision in isolation.

Increasingly, people are turning to computer systems for help in unravelling the complexities of modern business life. The past decade has seen the meteoric rise of the pesonal computer. Its raw processing power coupled with the sophistication of its applications has enabled more people than ever before to make wide use of models to support decision making. These models are now fed by real-time access to a huge variety of data sources made available through information superhighways and client/server computing. Modern information systems can show change as it happens, analyse the impact of that change and make recommendations as to action.

The increasing use of artificial intelligence techniques has made possible the use of systems which encapsulate expert knowledge, enabling inexperienced people to make high-quality decisions. Knowledge discovery techniques have been developed whereby huge quantities of raw data can be 'mined' for the information that they contain. Once the information has been successfully extracted it may then be used as the basis for future decision making.

In bringing together this book we have kept two goals in mind. Firstly, the goal of educating the reader by giving an insight into the wealth of computing and mathematical techniques now being used to build decision support systems. Secondly, we have aimed to stimulate the imagination by including an eclectic mix of contributions from a wide range of business areas to demonstrate that there is no field in which modern decision support techniques cannot usefully be applied.

To this end we have supplied an introductory chapter that examines some of the underlying reasons for the upsurge of interest in computer aided decision support, discusses the major principles and key technologies, and looks at future trends and developments. The next three chapters look at the important technology of data mining — the conversion of large volumes of data into high-value information for decision making. Chapter 2 by Shortland and Scarfe introduces the topic by considering case studies of a number of applications that have been investigated at BT Laboratories over the past few years. In Chapter 3, Limb and Meggs provide an overview of the many techniques being used for data analysis. To conclude the data mining trilogy, Tattersall and Limb consider, in Chapter 4, ways in which complex multi-dimensional data can be compressed to enable visualization and data exploration.

One of the key decision-making areas for any business is investment appraisal. Davies and Napier, in Chapter 5, describe a computer supported aproach that has been developed for use in BT. The important element of their approach is the modelling of the whole-life costs of an investment decision and not simply the up-front purchase cost.

It has already been stressed that we live in an age where we are surrounded by huge quantities of information of an increasingly unstructured form. For example, there surely cannot be a discipline which is not represented in an Internet news group, bulletin board or WorldWideWeb server. Revett and Benyon review, in Chapter 6, the importance of information of all types to decision makers and consider some of the issues arising out of modern information systems techniques. In particular, they examine the growing importance of unstructured information sources and the application of hypertext techniques.

The next group of five chapters has a strong telecommunications flavour. They describe in some detail how the principles and techniques of computer aided decision are being applied by BT in the areas of configuration management, contract specification, service provisioning, flexible bill design and network upgrade planning. Scott and Bull, in Chapter 7, describe a configuration and planning system implemented to provide support to systems engineers responsible for BT's ServiceView™ network management system. Their approach makes extensive use of graphical user interfaces and expert systems technology to reduce a complex design problem to one in which optimal solutions can be produced quickly and effectively. Chapter 8 by Last and Corley describes a new approach to contract specification in which case-based reasoning is used to advise negotiators of the most appropriate measures for managing a specific contract. This is a good example of decision technology being used to capture previous experience and reapplying it in new business situations. In a similar vein, O'Brien and Davison, in Chapter 9, describe a system to support service providers in selling their offerings over complex telecommunications networks. In particular they examine the use of fuzzy systems in multi-criteria decision making. Utton and Sanders provide us with an insight, in Chapter 10, into the complexities of bill design at a time when companies are having to respond to a rapidly changing market-place. The basis of their approach is a system that contains many hundreds of formatting rules. This allows the bill designer freedom to make changes while the system monitors the end result for appropriate information content and clarity. In the final chapter in this section, Crabtree and Munaf consider the problem of planning upgrades to very large telecommunications networks in the most cost-effective manner. The characteristics of this problem are shared by many other similar problems from other disciplines. In this case, the authors describe a solution using the technique of 'simulated annealing'.

The last contribution to this book, Chapter 12, looks at the increasingly important area of group decision making. O'Donnell shows that, at a time when companies are placing increasing emphasis on people working as teams, albeit in an environment of down-sizing, de-layering and dispersion, the advent of sophisticated group decision making techniques could well become vital to corporate survival. The chapter gives a thorough review of current group decision-making systems and draws some conclusions regarding future trends.

To pursue the metaphor outlined at the beginning of this preface it is clear that we are today still in a war. The protagonists are now the communications companies, the software developers and the information providers. The battlefield is the market-place. I hope that through this book you will see that the instruments of war have also changed and that the future belongs to those who most successfully apply decision-support techniques to the business problems of the day.

Finally, we would like to thank the many people who have contributed to the production of this book. In particular we should like to express thanks to our external reviewers, Professors Abe Mamdani of Queen Mary and Westfield College, London, and Nihal Sinnadurai now at Middlesex University, Dr Mike Uschold of the AI Applications Institute at Edinburgh, Graham Tattersall of UEA, Professor Alan Pearman of the University of Leeds, and Dr Roger James of Glaxo Wellcome R&D. We also wish to acknowledge both our many colleagues at BT Laboratories, in particular Professor Peter Cochrane, for their help, encouragement and advice during the production of this book, and the authors, without whom this book would not have been possible.

Phil Flavin and Ken Totton
BT Laboratories

1

AN OVERVIEW OF COMPUTER AIDED DECISION SUPPORT

K A E Totton and P G Flavin

1.1   INTRODUCTION

Computer aided decision support is of increasing significance to business at all levels. The effective exploitation of computer technology in the decision-making process is no longer the preserve of specialists, but a matter of corporate survival in a world of competition. Computer aided decision support draws upon recent advances in the fields of computing, management science, operational research, information management, and the psychology of human/computer interaction. Undoubtedly the greatest impetus has come from the availability of powerful low-cost personal computers (PCs). The resultant development of low-cost decision-support tools, such as the ubiquitous spreadsheet, together with the increasing availability of local and wide area communications networks has accelerated the penetration of the technology. All of this has led to an immense amount of information being brought to the desktop, allowing advanced computer aided support to be provided to decision makers.

The purpose of this chapter is to introduce the subject of computer aided decision support and to set the scene for subsequent chapters. In so doing, it is hoped that the diversity, ubiquity and richness of the subject will be conveyed to the reader. Whilst much of the literature emphasises managerial decision making, the treatment throughout this book will demonstrate that scope exists for the deployment of decision-support systems in all organizational systems at all organizational levels, from the boardroom to the shop floor.

For those who seek a thorough introduction to classical decision-support, some excellent texts are available. Turban's work [1] on the use of expert systems for decision-support lays emphasis on the managerial perspective. Klein and Methlie [2] consider theories of decision making, before focusing on the role of artificial intelligence (AI). Sprague and Watson [3] have edited a compendium of papers covering such topics as development, architecture and environments for decision-support systems (DSS), with chapters on executive information systems and group decision-support systems (GDSS). A recent contribution by Rhodes [4] emphasises the need to link theory to practice and follows two case studies.

The remainder of this chapter is set out as follows. Firstly, some key business imperatives are discussed in order to motivate interest in this topic. A conceptual framework for decision support is then presented, followed by the typical components of a decision-support system. Next, technologies suited to decision-support applications are introduced, followed by a discussion of some of the practical issues that arise. The chapter concludes with a short discussion of future trends.

1.2   BACKGROUND

The last two decades have witnessed a remarkable growth in the deployment of decision-support systems. High-quality decision making is now a key differentiator between those businesses that will survive and grow and those that will not. Some of the key business imperatives behind this upsurge of interest are discussed here.

The sheer diversity of business interests in today's market-place, coupled with the high degree of interdependence of business activities, means that no significant decision can be taken in isolation. Intelligence about the market, the competition, the political environment, and even the weather has a critical impact on decision making. Consequently, information is a key corporate asset. Access to information and its effective exploitation is a crucial source of competitive advantage. Without accurate and timely information it would not be possible to take rational decisions in today's highly complex world.

The inexorable drive to higher productivity and decreasing product life cycles accentuate the need for innovative, as opposed to merely efficient decision making. Business process re-design [5] is transforming entire corporations, requiring decision-support tools to manage effective change.

Additionally, the current trend for companies to 'down-size' is taking organizations to the point where their corporate knowledge is being eroded. In addition the public, customer and environmental impact on each company's operations is increasing. Today, companies have to pay attention to their 'outface', which results in even more information — as company information moves to enterprise knowledge.

In some cases, decisions can be taken by detailed analysis of a problem from first principles, provided that a sufficiently sophisticated model exists. More generally, decisions are based on experience, using a case history of previous situations as the basis for judgement. In both cases, either lack of technical skills or lack of previous experience threatens the general quality of the corporate decision-making process, hence generating keen interest in computer aided systems. This interest is often expressed in terms of consistency with respect to available information, as well as timeliness and objectivity. The latter point is considered to be essential in applications such as personal credit assessment (see Chapter 2) where justification of the decision may be required by law.

By any standards, the growth in computer technology in recent decades has been startling. Networking and, in particular, client-server computing have combined to greatly increase processing power and accessibility of data for decision making. The growth of corporate databases and the emergence of global information networks, such as Internet (see Chapter 6), have meant that, increasingly, decision makers have access to an invaluable repository of data on which to base their decisions. Advances in human/computer interfaces, especially the emergence of the graphical user interface, have brought powerful decision-support tools to the fingertips of the non-specialist.

1.3   A FRAMEWORK FOR DECISION SUPPORT

The demarcations between decision-support systems (DSS), and the related management information systems (MIS) [1] and executive information systems (EIS) are difficult to define. MIS and EIS often form part of the decision making process and generally share many technological components with DSS. Indeed these tools often share the same desktop computing platform. Because of this area of overlapping interest, it is useful to present a simple framework that provides the major concepts of decision support and addresses the relationship to other techniques of management science.

Simon [6] has shown that decision-making processes fall along a continuum that ranges from the highly structured (or programmed) to the highly unstructured (or non-programmed). Structured processes refer to routine and repetitive problems for which standard solutions exist, whereas unstructured processes are, by their nature, fuzzy and ill-defined and hence have no standardized solutions. Classically, the domain of decision support has been defined by Keen and Scott-Morton [7] as follows:

> 'Decision support couples the intellectual resources of individuals with the capabilities of the computer to improve the quality of decisions. It is a computer aided support system for management decision makers who deal with semi-structured problems'.

This definition, written in 1978, could be considered too narrow in the context of today's information technology. As will be seen throughout this theme issue, the domain of decision support is broad and capable of supporting the solution of both semi-structured and unstructured problems. In simple terms the relationship between EIS, MIS and DSS may be depicted by the triangle in Fig. 1.1.

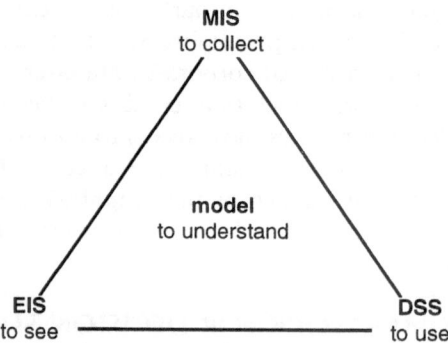

Fig. 1.1   MIS, DSS and EIS.

It is increasingly moving out of the realm of management science into many diverse facets of human activity. One of the key concepts underlying DSS, is that they incorporate both data and models. The prime purpose of modelling is to cope with data — to filter the superfluous and amplify the important. MIS and EIS solutions focus on data collation and the user interface aspects of information presentation. They make no real attempt to transform the data into the context of the external environment. This is left to the user of the information. A decision support tool will generally incorporate a mathematical model representing the problem on to which the data is mapped. This enables the synthesis of potential solutions to problems, and the representation of the impact they will have. Automated searching strategies can be employed to optimize potential solutions using a defined range of output conditions as a goal. Since most key decisions are made the basis of incomplete or ambiguous information, DSS should help people to think — properly developed systems help users make the correct deductions.

The decision support features introduced above may be summarized as follows:

- incorporate both data and models;

- assist in semi-structured and unstructured problem solving;

- support and not supplant human decision making;

- improve the effectiveness of decision making and not just the efficiency with which decisions are made.

Turban [1] has combined the taxonomy proposed by Anthony [8] with the work of Simon [6] in a decision support framework (Fig. 1.2). This framework is set against the context of management decision making and considers the type of management control exercised and the degree of structure in the problem domain.

The far righthand column of Fig. 1.2 indicates the type of support system that would typically be needed for decision making for a given class of

type of decision	type of control			support system
	operational control	managerial control	strategic planning	
structured	accounts receivable, order entry	budget analysis short term forecasting, personnel reports, make or buy analysis	financial management, distribution systems	MIS, operations research (OR) models, transaction processing
semi-structured	production scheduling, inventory control	credit evaluation, budget preparation, project scheduling	building new plant, mergers and acquisitions, compensation planning	DSS, 'soft OR' models (see section 1.5.1)
unstructured	buying software, approving loans	recruitment, negotiation	R&D planning, new technology development	DSS, 'soft OR' models (see section 1.5.1)

Fig. 1.2   A simple decision support framework (after Turban [1].

problem. The examples are merely representative and are not intended to be complete.

In the remainder of section 1.3, a general problem-solving strategy is introduced, followed by a discussion of the important topics of information access and decision-support models.

1.3.1   Problem solving

Problem solving is an activity almost universal in its application and consequently a number of problem-solving techniques have evolved to assist this process. This section introduces one of the best known and relates it to a simple three-stage decision process.

The steps in solving unstructured problems may be conveniently considered with reference to the total quality management (TQM) problem-solving process widely used within BT (Fig. 1.3).

As the wheel suggests, the process is usually iterative, and may involve several cycles of refinement until the desired solution is obtained.

I   Identify the problem — as with all problem solving, the task must begin with problem identification. This may be non-trival in that what appears to be a problem may be symptomatic of something more fundamental. A problem, or indeed an opportunity, is usually detected by reference to some business metric or goal.

II Gather data — for most problems it is necessary to gather historical or current data of relevance to the problem. Where data is unavailable (for example in situations with a high degree of novelty), estimation, assumption, or the use of more loosely related information may be necessary.

III Analyse the data — this step should establish the root causes of the problem. From this it will be possible to verify that a problem exists, estimate its significance, location, and establish ownership. Additionally, classification may be used here to assign the problem to some definable category so that previous cases may be drawn upon during the next step. As mentioned above, the degree of problem structure is a key criterion during this process.

IV Generate solutions — in many cases where there is a reasonable degree of structure to the problem, case history may be used to identify possible solutions. Computer-based techniques lend themselves readily to problems in this class. In ill-structured problems, techniques such as brainstorming can be applied effectively to obtain novel alternatives. For

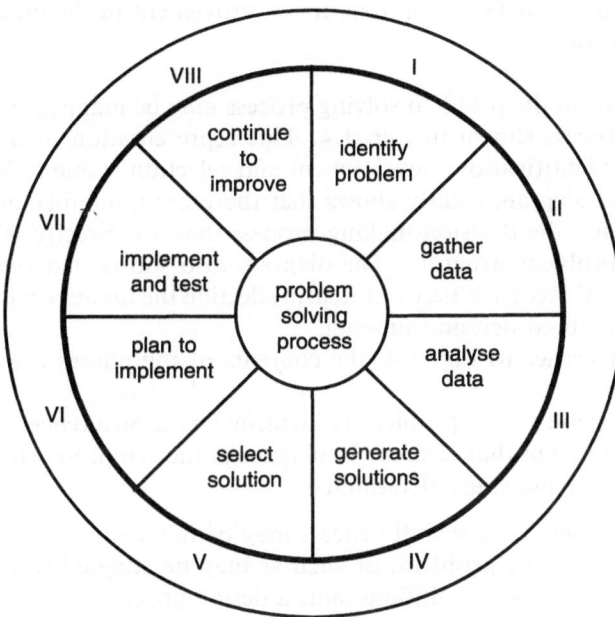

Fig. 1.3 TQM problem-solving process.

each set of alternatives identified, their impact must be evaluated. Computer aided techniques are used to build models of solutions, or to populate stock models with problem-specific data. There can be a high degree of iteration at this point in order to obtain an optimized solution and a number of advanced computing techniques have been developed for this purpose, particularly in the AI domain.

V   Select the solution — the evaluation of alternatives depends on the criteria selected for solution acceptance. In some cases the optimum solution may be unattainable, but one might be prepared to settle for a satisfactory solution, usually involving much less computation. This process is known as 'satisficing' [6].

VI Plan the implementation — the solutions developed need to be carefully considered in relation to their implementation in the real world. Human and organizational factors will play a crucial role in determining the success of the chosen solution.

VII Implement and test — the key issue at this stage is to determine whether the problem has been solved. This requires suitable metrics for monitoring and performance evaluation.

VIII Continue to improve — given that the solution has been implemented, this stage reviews the options for improvement in the effectiveness of the solution.

The steps in the problem solving process may be mapped directly to the decision process shown in Fig. 1.4. This representation of a three-phase process of identification, development and selection is due to Mintzberg et al [9]. The diagram clearly shows that there are a number of alternative routes though the decision-making process that are directly related to the degree of problem structure. The diagram also shows that there can be a high degree of feedback between steps, reflecting the iterative nature of semi- and un-structured decision making.

With reference to Fig. 1.4, the contents of the phases are as follows.

- Identification — a problem is identified as a difference of the actual situation from that desired. In diagnosis, the symptoms are sorted and the underlying cause determined.

- Development — a search process may be initiated to find ready-made solutions to the problem, or such as may be adapted to the problem. Should no existing solutions suit, a design process must be invoked to develop a custom solution.

- Selection — in situations where many possible solutions are identified, a screening process may be used to examine them in depth and discard

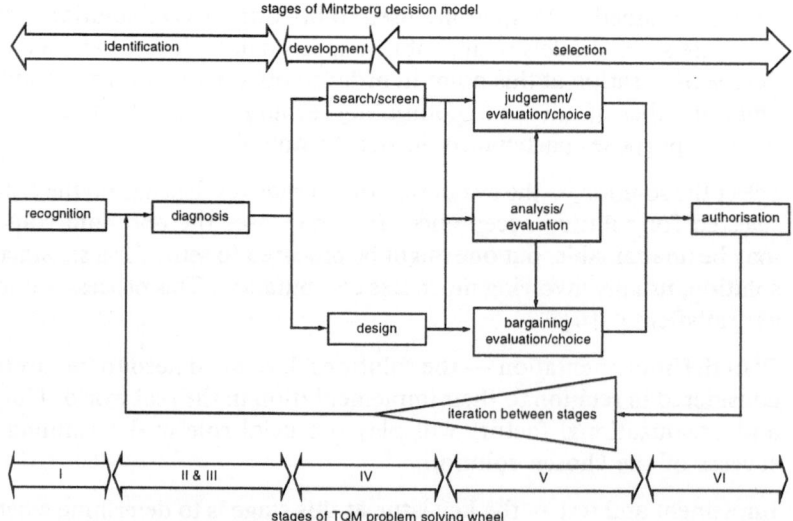

Fig. 1.4   A model of the decision process (after Mintzberg et al [9]).

the non-feasible. Evaluation and choice may proceed in three modes, typically dependent on the degree of structure. Judgement is often exercised in ill-structured problems. Analysis and choice on the basis of objective criteria (e.g. value) is possible in structured problems. Finally, authorization commits the organization to enacting the selected course of action [2].

1.3.2 Information access

Convenient access to information is central to decision making and a vital component of Mintzberg's identification stage. Indeed given effective presentation of all the information relevant to a problem, the decision can often be straightforward. Since corporate information is a key asset, it is increasingly being extracted from operational systems, stored, and managed within large data warehouses [10]. The bringing together of large amounts of data in this way facilitates the development of data mining techniques (see Chapters 3 and 4), enabling underlying trends and patterns to be observed directly or to be inferred using the techniques of machine learning (see Chapter 2).

It has already been indicated that direct access to global information repositories is becoming a reality. In Chapter 6, Revett and Benyon cover this topic in some detail, and emphasize the advantages to be gained from intelligent access to corporate and external information sources. They address the infrastructure in terms of networking and standards necessary to provide easy access to corporate information. Based on such recent phenomena as the growth of the Internet, they contend that the development of hypertext and hypermedia based information systems, integrated with office automation and relational databases, will allow users to access and browse a vast range of otherwise inaccessible information.

1.3.3 Models

Decision support systems are required to model elements of the real world to promote understanding and to assist the decision process. It is not possible, and often not necessary, to model the real world to an arbitrary level of detail. Model design is, however, a critical activity for the DSS developer, a key objective being to extract the features of interest. The benefits of developing models include the following:

- filter — allow generalization from specific examples avoiding detail;

- amplify — support prediction of known cases and archetypes;

- argue — provide for explanation and 'what-if' analysis.

Modelling involves analysing the problem and abstracting it into suitable symbolical or numerical terms, and trade-offs exist between the precision of the model and its complexity. Modelling a complex decision domain is a major undertaking. Some of the modelling concepts frequently encountered in the literature are listed below.

- Variables and their mathematical relationships must be determined. 'Decision variables' represent quantities over which the decision maker has control. External variables will be 'givens' associated with the environment, e.g. interest rates. Result variables will reflect the effectiveness of the system.

- Searching is required to find a solution to the model, i.e. a set of values for the decision variables. Several search strategies are possible:

  — algorithmic techniques increase search efficiency, being a set of steps towards an optimal solution (if it exists);

  — blind search which simply enumerates alternatives until a satisfactory solution is found, suffers from combinatorial explosion and is usually impractical on large problems;

  — heuristic search exploits decision rules, e.g. 'rules of thumb' about how a problem should be solved. There has been much interest in this area in recent years, the focus being on expert systems that seek to encapsulate knowledge about the decision-making process. In recent years the techniques of artificial intelligence such as expert systems, constraint logic programming, genetic algorithms, neural networks and simulated annealing (see Chapter 11) have been successfully applied to this task (see section 1.4).

- Sensitivity analysis is often carried out once a solution to the model has been obtained. This tests the robustness of the solution and the underlying model in the face of likely uncertainty in the external variables, and possible interactions between variables. Sensitivity analysis is an automatic by-product of many techniques, for example linear programming [11].

- 'What-if' analysis is a trial-and-error method. It determines the impact on the solution of changes in decision variables and assumptions. With modern graphical user interfaces, this type of analysis is very easily performed, e.g. by adjusting a slider representing the value of a variable.

- Goal seeking is concerned with determining the values which must be assigned to the input variables if a particular output is to be obtained, e.g. what growth in call volumes must be obtained in order to increase annual profits by 10%.

- A scenario defines the variables and parameters in which a decision situation is to be examined. Scenarios are important in decision support, since they describe the setting in which the decision is to be evaluated, i.e. the decision variables and 'givens' for a particular situation which is to be modelled. Multiple scenarios may well need to be evaluated to account for uncertainties in a rapidly changing environment.

Modelling complex decision domains inevitably requires careful model management. Typically, model management involves the following phases.

- Model creation — the formulation, implementation and validation of models. There may be scope for reuse of previously developed models. Composition involves the linkage of models such that the output of one becomes the input of another. Integration may require changes to the models, e.g. where the models exist at different granularities. Finally, formulation is the process of converting a precise problem description into a mathematical model.

- Using the models — running the models and interpreting the results.

- Maintenance — including conversion of complex models into simple heuristics.

In any practical situation there will be iteration until a satisfactory outcome is achieved. Some common modelling technologies will be briefly mentioned in section 1.5.

1.4   GROUP DECISION-SUPPORT SYSTEMS

No discussion of computer-aided decision support would be complete without reference to group decision-support systems (GDSS). They define an important sub-field within decision support, reflecting the fact that most important decision making in organizations is undertaken not by individuals, but by a team. Indeed, a recent survey of BT managers indicated that 70% felt that decisions were made by groups. GDSS are a form of groupware [12], reflecting the increasing importance of the business team, and fuelled by technology developments such as the reducing costs of networked computing, communications, and work-group applications such as Lotus Notes ™. In addition, several business issues are combining to focus attention on group support:

- effectiveness — down-sizing of companies, de-layering of organizations, dispersion of work-forces;

- timeliness — need for rapid response;

- location — need to reduce amount of business travel;

- reach — transience of work-groups and companies;

- interdependence — no single individual, or indeed organization, 'knows it all'.

In Chapter 12 O'Donnell addresses some of the current thinking in the area of group decision making. This is an area which in the past has promised much but which has delivered little, due largely to inflexibility and lack of perceived value to organizations. O'Donnell emphasizes the importance of group dynamics in group decision making, and then introduces the decision environments available in terms of computer hardware, software, communications, and procedures. The chapter then surveys some of the GDSS described in the literature, and concludes with a list of key considerations which GDSS developers need to bear in mind.

Further discussion of this topic is beyond the scope of this introductory chapter, but the interested reader will find several papers on GDSS in Sprague and Watson [3].

1.5 DSS TECHNOLOGIES

This section covers some of the techniques and technologies which are being used to build decision-support systems. Following a summary of modelling techniques, an overview of relevant AI techniques is presented.

1.5.1 Modelling techniques

Operations research is concerned with the mathematical modelling of systems or problems, generally with the aim of identifying an optimal solution. The field of operations research includes a number of techniques of use to decision-support systems. For example, a linear programming approach may be adopted to find the optimal solution where a number of variables apply; queuing theory can be used for the mathematical simulation of some dynamical systems. Recently, 'soft OR' has emerged as a sub-field characterized by much less reference to mathematical or statistical techniques, and more concern with ways to structure and understand problems [13].

Highly relevant to decision support systems is the field of decision theory (which has its roots in operations research) and its practical application in decision analysis. Decision theory and analysis can be used to model a completely rational decision maker (as opposed to the decision-making domain), to help ensure that the decision is consistent and to understand the effect of external and environmental factors. It does not replace the expertise or knowledge of the decision maker, as AI techniques can, but improves the robustness of the decision-making process. It does this by using modelling techniques such as multi-attribute utility theory [14]. For information on the techniques of decision theory and decision analysis the reader is referred to French [15] and Watson and Buede [16]. Work on the psychology of decision making, highlighting some of the irrational or inconsistent processes sometimes used by decision makers (including experts) is described in Kahnemann, Slovic and Tursky [17].

One modelling option that is increasingly employed is that of object oriented modelling [18]. The object oriented paradigm offers the powerful mechanisms of encapsulation, abstraction, and polymorphism. This permits an evolutionary approach to modelling, and facilitates reuse of models. Classes and objects are directly mapped to the real-world entities that they model. Powerful simulations can be carried out since methods controlling the behaviour of objects are encapsulated within the object definition.

1.5.2   Artificial intelligence

In recent years there has been much interest in the application of artificial intelligence techniques to decision support. Rich [19] has defined AI as 'the study of how to make computers do things which, at the moment, people do better.' In the context of decision support this could, perhaps, be paraphrased as 'the study of how to make computers support people in producing good decisions where at present they make average ones unaided'.

Central to any discussion on AI is the question of the definition of intelligence. For the purposes of this chapter an intelligent system is considered to be one which exhibits one or more of the following characteristics:

- assimilates and exploits 'domain' knowledge in order to solve problems;
- uses general problem-solving capabilities;

- draws on experience of similar cases;

- reasons about a problem and generates new facts by inference;

- adapts or learns through a process of training.

A number of AI technologies have been applied to the development of intelligent decision support systems. Perhaps the most notable impact has been from the use of expert or knowledge-based systems. Henrion et al [20] offer a comprehensive treatment of the relationship between the fields of decision analysis and expert systems. Recently there has also been widespread interest in machine learning. These topics are outlined in the following subsections.

1.5.2.1 Knowledge-based systems

Knowledge-based systems (KBS) incorporate knowledge of the problem domain. In a KBS, this knowledge is typically separated from the mechanisms needed to process the knowledge (the inference engine). This modularity enables developers to concentrate on modelling the problem domain, and should provide for easier maintenance. Examples of the use of these techniques in service provisioning are given in Chapter 9, and in telephone bill design in Chapter 10.

Experts have at their disposal a set of heuristics and short cuts that they use to improve their performance. This knowledge is sometimes known as shallow knowledge. Rule-based systems encode knowledge as a set of rules or heuristics. An inference engine can chain the rules to map a set of symptoms to possible causes, a classical example being fault diagnosis. Maintainability of the rulebase in a changing environment can be a major issue for these systems, since a small change in the problem domain can necessitate costly re-engineering of the rule base. For an example of the application of this type of system in network management system design, and a strategy for rule maintainability, see Chapter 7.

As part of their education, experts learn about the theories, axioms, and laws of their domain of expertise. This deep knowledge allows them to reason from first principles in situations where their rules of thumb are not applicable. Model-based systems (sometimes known as deep modelling) exploit functional and structural models of the problem domain to facilitate problem solving. This means that the system in principle can be used to solve any problem to a level of detail commensurate with the underlying models. They offer advantages over rule-based systems in terms of generality and the fact that the models may be represented graphically. The latter feature is an important aid to user comprehension.

Case-based reasoning mimics the human capacity for problem solving by reference to a set of previous cases [21]. The parameters of a new problem are compared, using a fuzzy-matching algorithm, with other examples held in the case base. Those most closely matching are used as the basis of a solution to the new problem. This approach holds out the prospect of more maintainable systems, as, in principle, a new case may be added to the case base without the large-scale revisions often required by rule-based systems under similar circumstances. For an example of the application of case-based reasoning to contract specification, see Chapter 8.

1.5.2.2   Machine learning

In recent years, it has become apparent that, through the penetration of information technology, the ability to store and manipulate information has far outstripped the ability to analyse it. For example, in pharmaceutical research and development, data productivity has been increasing at 20% compound since the early 1970s yet new products only grew at 5% over that period. This has accentuated the need for techniques which can be used to distil knowledge from large volumes of data. In many cases, access to relevant data may be possible, enabling the induction of models directly from the data. This can augment or replace the need for the elicitation of expertise from a human expert. The ability to learn facts directly from data or adapt to new situations is an important facet of an intelligent system. The overall field, known as machine learning, finds its most important application in decision support as a pattern classification technique. In many applications, trends or patterns in large volumes of data hold the key to good decision making. All of the techniques described here provide a useful insight for the DSS user.

Inspired by the biological structure of the brain, neural networks comprise large arrays of simple processing elements (artificial neurons), connected by weighted links. In multilayer networks, each neuron is, typically, fully connected to all the neurons in the adjacent layer. Training takes place by the successive presentation of patterns at the network inputs and associated outcome class at the outputs. During repeated passes through the training set, the link weights are adjusted until they converge to their final values, at which point the network is deemed to be trained. Thereafter, new data sets may be classified using the trained network with a high degree of accuracy providing that the new data and the training data belong to similar populations.

Decision tree and rule induction constitute a well-established set of techniques which use a process of symbolic induction. The ID3 family of algorithms is perhaps the most widely utilized [22]. Using ideas drawn from information theory, the algorithms recursively select the parameters that most effectively split the data set into the various outcome classes. The resultant decision tree is an explicit representation of the knowledge implicit in the data set, and can be used directly for decision making or converted to a set of rules for building an expert system.

In addition to neural networks and induction techniques, statistical pattern classification is also important to the field [23]. Some of the more important techniques are regression, clustering, and nearest neighbour classification [24].

These techniques are increasingly being used in combination to develop powerful decision support aids. For example, a clustering algorithm might be used to identify clusters in the data; decision-tree induction could then be applied to identify the distinguishing attributes of the various clusters. Assisted by data visualization techniques [25], the discovery of new knowledege from large databases is known as data mining. The underlying techniques are treated in Chapters 3 and 4. Applications in areas such as customer account management and marketing are discussed in Chapter 2. The current trend towards data warehouses provides a rich platform for the effective application of data mining techniques [26].

1.6 BARRIERS TO SUCCESSFUL DSS DEVELOPMENT

The successful development and exploitation of computer aided decision support must overcome several barriers. The issues discussed here are by no means unique to DSS but are nevertheless crucial to successful development and operation.

One of the most challenging aspects of decision support application development is knowledge capture. The issues here include:

- knowledge of what to look for (problem identification);
- identification of domain experts;
- gaining access to experts;
- handling disagreement among experts;
- lack of mature models describing the problem domain;
- access to adequate historical data for machine learning;
- rate of change in the problem domain;
- inclusion of commonsense (the standard AI problem — 'the dead patient is cold' — usually results in the advice to give him a blanket!)

Real world decision domains typically exhibit rapid change. This severely challenges the derivation, validation, and especially, the maintenance of complex models. For example, in the credit assessment work reported in Chapter 2, the basic assumption is that customer behaviour in the recent past is a reliable predictor of future behaviour [27]. It follows, therefore, that any particular model of customer behaviour can have only a limited life expectancy. For all but the most *ad hoc*, 'throw-away' applications, the DSS developer must design the system to cater for change. On the other hand, in many cases, although individuals change their behaviour they are changing between well-known archetypes, e.g. a patient with a progressive chronic disease may change condition from 'mild' to 'severe'. Reuse of models may also be an important consideration. In the longer term there is the prospect of adaptive systems capable of tracking changes in the decision environment.

By definition, decision support involves the decision maker. Thus, it is to be distinguished from task automation in that the user is an integral part of the overall process. This is important for the following reasons:

- any system's world model will be imperfect — due allowance may need to be made for factors relevant to the problem which are not modelled in the system;

- users will need to be aware of the limitations (e.g. lack of completeness, or expressive power) of the various models;

- solutions to complex problems are typically iterative in nature, with the user performing 'what-if' analysis, modifying parameters and steering the overall search process;

- human judgement will be necessary to select among solution options where factors such as political considerations are important and where the perspective on the problem can vary, say from regulator to consumer.

In some situations, acceptance of computers may be a threat to an otherwise excellent system. As acceptance of information technology increases, this is likely to become less of an issue, but attention to the following should reduce this risk:

- adoption of a user-centred design approach;

- use of the technology as a 'thinking amplifier', not a replacement thinker;

- involvement of users at each step of the development process;

- promotion of usability, ensuring conformance to all relevant user interface standards.

System flexibility and interaction design are also key, in order to stimulate the user to investigate multiple-decision scenarios.

1.7 FUTURE TRENDS

The continuing fall in price of personal computing linked with the rapid expansion of access to information will continue to fuel the development of advanced decision-making tools. Already, large corporate data warehouses are being built, which draw together data from disparate operational systems into a form that can be readily accessed and exploited. New trends are emerging such as Ringo (the MIT home page — http://www.media.mit.edu) which collects individuals' trends and opinions on music in order to bring together a community of experts each identified by their data contributions. This trend will undoubtedly continue and will be supported by the development of a myriad of decision-support tools accessing this data for a wide variety of purposes. For large corporations, these systems will become vital to business success. It is becoming widely accepted that, as companies change their size and structure, they will no longer be able to afford paper-based transactions and reporting. The availability of accurate information coupled with computerized process models will enable rapid and accurate decisions to be made, increasing their profitability while reducing their costs.

As artificial intelligence technology continues to mature, more intelligent systems will be developed. As discussed in section 1.5.2, machine learning technology can highlight trends in data that would not be visible using other techniques. On the basis of this approach, systems that adapt their behaviour because of changing circumstances are already beginning to emerge (e.g. automated trading). Systems that can co-operate across data networks to solve problems or which can gradually evolve to improve on historical solutions are already reported [28]. The impact of these approaches is likely to be profound. Whereas today's decision support tools tend to use fixed models to describe the domain in which they are operating, the emergence of systems with more general problem-solving capabilities is likely.

An example of this trend is the advent of the personal digital assistant (PDA). Today's PDA is a cleverly packaged computer with some sophisticated user interface and communications capabilities (handwriting recognition, infra-red links, etc). As the infrastructure to support PDA use matures, the availability of on-line information from anywhere in the world coupled with distributed intelligent processing will enable really sophisticated decision support applications to be developed. For example, it will be possible for purchasing decisions to be made by instant look-up of prices or whole-life costs (see Chapter 5) on almost any product. Obvious examples are the purchasing of cars, houses or holidays. Financial planning could be transformed as support tools offer completely impartial advice on investment and taxation. Planning and scheduling activities, such as diary management and appointment negotiation could become tasks that are fully delegated to a user's PDA.

Another trend, facilitated by the modern communications infrastructure, is for organizations to become more widely distributed and to employ fewer full-time workers. The time, costs and inconvenience of modern travel is a significant barrier to face-to-face meetings and hence the use of low-cost videoconference technology is growing. Also growing is the use of news groups and similar facilities on networks such as Internet where people with shared interests can communicate and exchange ideas without ever having met in person. The growth of electronic communities of this sort is bound to continue. As a result, interest in group decision making and decision conferencing is growing. The developments required are as much to do with organizational development, psychology and group dynamics as they are with technology. The essential role of the technology is firstly to make these new insights possible, but secondly to allow the users direct access to their problems and their data — the technology has to ensure it does not get in the way.

1.8   CONCLUDING REMARKS

This opening chapter has introduced the field of computer aided decision support and outlined its breadth and importance to modern businesses at all levels. The importance of decision-support technology can only grow as the impact of modern communications expands the reach of both individuals and corporations. It sets the scene for the more detailed treatment of specific applications in this book. Hopefully, the reader will have gained a good insight into the potential for deployment of decision-support technology, both now and in the future, and will be encouraged to read further.

APPENDIX

List of abbreviations

AI   artificial intelligence
DSS   decision-support system
EIS   executive information system
GDSS   group decision-support system
KBS   knowledge-based system
MIS   management information system
OR   operations research
PC   personal computer
PDA   personal digital assistant
TQM   total quality management

REFERENCES

1. Turban E: 'Decision support and expert systems', MacMillan (1988).

2. Klein M and Methlie L B: 'Expert systems: a decision support approach', Addison Wesley (1990).

3. Sprague R H and Watson H J: 'Decision support systems — putting theory into practice', Prentice-Hall International (1986).

4. Rhodes P C: 'Decision support systems: theory and practice', Alfred Waller (1993).

5. Harrington H J: 'Business process improvement', McGraw-Hill (1991).

6. Simon H: 'The new science of management decision', New York, Harper and Row (1960).

7. Keen P G W and Scott-Morton M S: 'Decision support systems, an organizational perspective', Reading MA, Addison-Wesley (1978).

8. Anthony R N: 'Planning and control systems: a framework for analysis', Harvard University Graduate School of Business, Cambridge MA (1965).

9. Mintzberg H et al: 'The structure of the unstructured decision processes', Administrative Science Quarterly, 21 , No 2 (June 1976).

10. Inmon W H: 'Building the data warehouse', QED Information Sciences (1992).

11. Bundy B D and Garside G R: 'Linear programming in Pascal', Edward Arnold (1987).

12. Rogers A S (Ed): 'Groupworking over networks', BT Technol J, 12 , No 3 (special issue) (July 1994).

13. Blattberg R C, Glazer R and Little J D C (Eds): 'The Marketing Information Revolution', Harvard Business School Press (1994).

14. Keeney R and Raiffa H: 'Decisions with multiple objectives: preferences and value trade-offs', Wiley, New York (1976).

15. French S: 'Decision theory', Ellis Horwood (1986).

16. Watson S and Buede D: 'Decision synthesis', Cambridge University Press (1987).

17. Kahnemann D, Slovic P and Tursky A: 'Judgement under uncertainty: heuristics and biases', Cambridge University Press (1982).

18. Cusack E L and Cordingley E S: 'Object orientation in communications engineering', BT Technol J, 11 , No 3, pp 9-17 (July 1993).

19. Rich E: 'Artificial intelligence', McGraw-Hill (1983).

20. Henrion M, Breese J S and Horvitz C: 'Decision analysis and expert systems', AI Magazine, pp 64-91 (Winter 1991).

21. Kolodner J L: 'Improving human decision making through case based decision aiding', AI Magazine, pp 52-68 (Summer 1991).

22. Quinlan J R: 'C4.5: programs for machine learning', Morgan Kaufmann Publishers, San Mateo CA (1993).

23. Gordon A D: 'Classification', Chapman & Hall, London (1981).

24. Chatfield C and Collins A J: 'Introduction to multivariate analysis', Chapman & Hall, London (1986).

25. Walker G et al: 'Visualization of telecommunications network data', BT Technol J, 11 , No 4, pp 54-63 (October 1993).

26. Rosenhead J: 'Into the Swamp — the analysis of social issues', J of the OR Soc, 43 , pp 293-305 (1992).

27. Lewis E M: 'An introduction to credit scoring', Athena Press (1992).

28. Koza J R: 'Genetic programming', MIT Press, Cambridge, MA (1992).

2

DATA MINING APPLICATIONS

R Shortland and R Scarfe

2.1 INTRODUCTION

'It has been estimated that the amount of information in the world doubles every 20 months. The size and number of databases probably increases even faster'.

W J Frawley et al [1]

With the increased use of computers, there is an ever-increasing volume of data being generated and stored. The sheer volume held in corporate databases is already too large for manual analysis and, as they grow, the problem is compounded. Furthermore, in many companies data is held only as a record or archive. BT has huge volumes of data from 20 million customer accounts, call records, equipment records and fault logs. Potentially valuable information is hidden within these databases and is under-exploited. As Sir John Harvey-Jones says: 'IT has failed to move from data processing to becoming a key strategic weapon to change businesses in ways to beat the competition. The real value of IT is only realized if you change the way business is done' [2].

This chapter presents a number of case studies demonstrating how data mining is being used to exploit valuable data.

Data mining encompasses a range of techniques which aim to create value from volume and form the foundation of decision making (Fig. 2.1). It does not have a formal definition and there are differing views on its meanings.

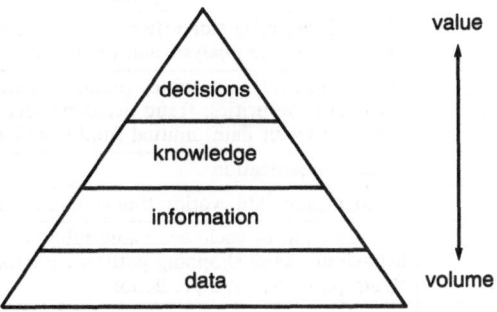

Fig. 2.1   From volume to value [3].

For the purpose of this chapter data mining is defined as the process of extracting implicit information from databases, often by using various computerized analysis techniques in combination. These are drawn from the disciplines of data analysis, machine learning (see Chapter 3), and data visualization [4]. They include cluster analysis, dimensional compression, neural networks and tree-and-rule induction.

The case studies presented in this chapter highlight the use of these techniques and the wealth of information contained in databases. The studies presented are in the following areas:

- identifying faults on printed circuit boards;

- discovering the organizational structure of groups of criminals;

- market segmentation and customer characterization;

- predicting outcomes of credit assessment and litigation.

2.2   APPLICATIONS

2.2.1   Background

By way of introduction it is worth considering the wide range of applications where data mining has been used. Table 2.1, derived from Frawley et al [1], is not intended as an exhaustive list but demonstrates the range and diversity of applications.

Table 2.1   Data mining applications.

Industry	Application areas
Medicine	biomedicine, drug side effects, hospital cost containment, genetic sequence analysis and prediction
Finance	credit approval, bankruptcy prediction, stock market prediction, securities, fraud detection, detection of unauthorised access to credit data, mutual fund selection
Agriculture	disease classification
Social	demographic data, voting trends, election results
Marketing and sales	identification of socio-economic subgroups showing unusual behaviour, retail shopping patterns, product analysis, frequent flying patterns, sales prediction
Insurance	detection of fraudulent and excessive claims, claims 'unbundling'
Engineering	automotive diagnostic expert systems, Hubble space telescope, computer aided design (CAD) databases, job estimates
Physics and chemistry	electrochemistry, superconductivity research
Military	intelligence analysis, data fusion and other classified applications
Law	tax and welfare fraud, fingerprint matching, recovery of stolen cars

2.2.2   Case study 1 — fault diagnosis

One of the earliest studies of data mining in BT employed neural networks for automatic diagnosis of faults in line cards used in digital switches [5]. Previous attempts using a 'deep model' expert system approach [6] were hampered by the time and expertise required to build and maintain an adequate model based on a knowledge of the circuit functions. Neural networks offered the possibility of automatically obtaining a 'shallow' model (implicit in the trained network) sufficiently detailed to diagnose the fault classes to the required level of accuracy. This model was based on readily available past experience in the form of previous test results and diagnosis outcomes held in repair databases.

Machine learning[1] classification techniques attempt to build rules that distinguish between classes. They require data to be presented as a set of attributes followed by the class to which the example belongs, as shown in Fig. 2.2. In the case of the fault diagnosis system, the test data was applied to the input of a neural network, which was then trained to recognize the

---

[1]Unless explicitly specified, in this chapter machine learning refers to that area of machine learning concerned with building classifiers.

appropriate fault class. By way of illustration, Fig. 2.3 shows six examples of test outcomes for two of the possible fault types. In practice, the training data consisted of about 250 example cases which had been classified using manual techniques.

Attr 1	Attr 2	Attr 3	Attr 4	Attr 5		Attr n	Class

Fig. 2.2   Attributes for classification.

Test 1	Test 2	Test 3	Test 4	Test 5		Test 77	Fault type
3.700	-3.700	-0.002	-1.914	4.000		?	Component 1
39.400	-39.00	-0.003	-2.054	0.160		7	Component 2
39.900	-39.10	-0.002	-0.518	0.160		7	Component 2
-7.400	-55.80	-0.002	-0.032	4.00		?	Component 1
38.300	-39.10	-0.001	-0.518	0.230		8	Component 2
2.100	-2.100	-0.002	-0.518	4.00		?	Component 1

Fig. 2.3   Test case data format.

One issue which arose in this case study was that of 'missing data'. The full test procedure involved 77 separate tests. In practice, to improve test machine throughput, the process was terminated once an abnormal test result was identified. Incomplete tests therefore resulted in values missing from the test set. A neural network expects an input value for each attribute and consequently incomplete test sequences pose a problem. Three ways of tackling this problem were considered:

- eliminating examples with missing values;

- generating a random number within the valid range for the attribute;

- using a fixed number at the mid-point of the valid range.

The first option was not viable as most examples have missing entries to some degree. Of the remaining two it turned out that using the mid-point of the valid range was the most successful [6].

A further lesson learned was that the best results could be obtained by limiting the number of fault classes. Initial experiments examining the ten fault classes were unable to achieve acceptable accuracy. However, analysis of the data revealed that the complexity of the initial task could be reduced significantly. By exploiting the fact that 85% of faults were due to only four component types (see Fig. 2.4), a classifier was built which achieved an overall classification accuracy of 92%. This was of equivalent accuracy to the 'deep model' approach, but provided much faster classification and offered an efficient 'first pass' diagnosis. The remaining 15% of faults could be diagnosed during a 'second pass' using the 'deep model'.

The ability to understand and modify the original task can be essential to achieve the best results from data mining. In this case reducing the complexity of the problem had a dramatic effect on the overall accuracy of the solution.

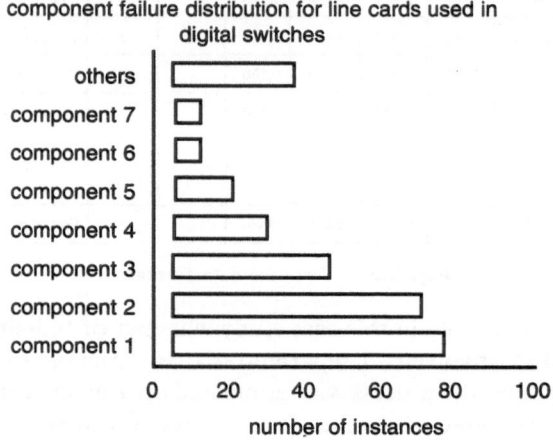

Fig. 2.4   Identifying problems.

2.2.3   Case study 2 — fraud

This case study shows how visualization can be used to identify relationships between entities in a database in order to support further exploration. Investigating a fraud perpetrated against the telephone network rapidly revealed features that suggest data mining is of primary interest:

- fraudsters are often part of a highly organized gang — implying that the data might have structure;

- many of those committing crime can be identified individually;

- very often extensive call records are associated with crime;

- sufficient data has to be available to identify the fraud and the fraudsters.

To assist in identifying a criminal hierarchy, a data visualization tool was used to identify calling patterns from a large number of call records. Figures 2.5 and 2.6 show telephones represented as nodes on the circumference of a circle and telephone calls as links between them. Figure 2.5 gives an impression of the complexity of the initial problem.

The approach taken was to explore the theories of security investigators. The most significant of these was that 'premium rate services targeted by the largest number of fraudsters are most likely to be part of an organized crime.'

This was explored by simplifying the display to show calls only to premium rate services of interest (as shown in Fig 2.6).

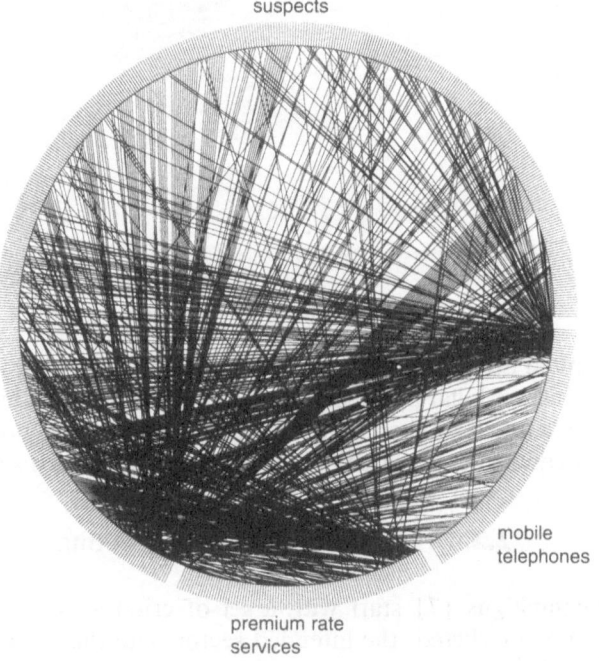

Fig. 2.5 Initial view of the fraud problem.

By exploring similar theories the investigators successively refined their understanding of how each of the fraudsters fitted into the criminals' organizational structure. Such information has allowed investigators to concentrate their effort on people who were most likely to be the ringleaders. This has resulted in a number of successful arrests.

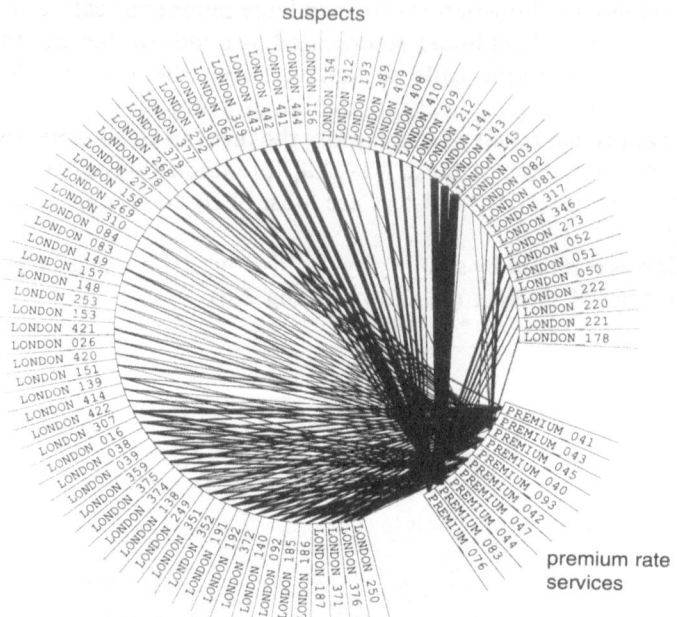

Fig. 2.6   Major fraud problem.

In the United States, where this type of fraud originated, it is estimated to be a multi-billion dollar problem. It is relatively new in the United Kingdom and occurs on a much smaller scale. By being proactive it is possible to limit and, in many cases, prevent or contain it from becoming such a problem here.

2.2.4   Case study 3 — marketing

Marketing campaigns [7] start with a set of criteria — for example, the product(s) to be marketed, the intended sector, and the geographical area. Machine learning (see Chapter 3) can be applied at the initial stage to characterize customers who already have the product. These characteristics can be incorporated into the campaign criteria to optimize the targeting and response rate (Fig 2.7).

As the campaign progresses, machine learning can be used to characterise those customers who respond positively and to refine the targeting for subsequent cycles of the campaign. This leads to a greater level of efficiency, achieved by targeting customers with a high probability of accepting the product.

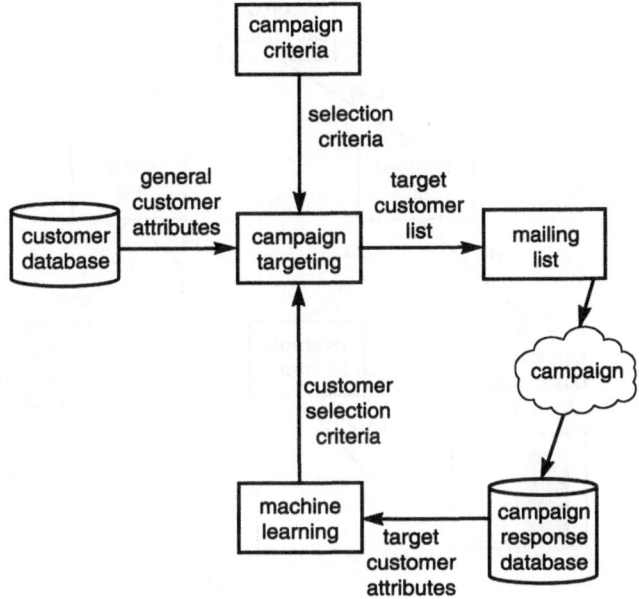

Fig. 2.7   Targeting customers.

The following case study shows how machine learning has been used to characterize customers. The training data set extracted 2000 customers who had opted for and against a raft of network services. Figure 2.8 shows part of a decision tree (see Chapter 3) produced by machine learning.

This work highlights the 'maximum call charge' as being the most significant indicator for determining whether or not a customer is likely to use network services.

The next most significant factor is the 'itemized bill' indicator. It can be seen that customers with itemized bills and call charges greater than £56 are very likely (83% probability) to use network services. Those who do not have itemised bills and have call charges less than £56 are unlikely (87% probability) to use network services.

The insights gained by the automatic generation of decision trees can be used in a number of ways — for targeting offers to people with similar characteristics, and to help understand why people choose or disregard a service.

Machine learning also offers the following benefits to support marketing campaigns:

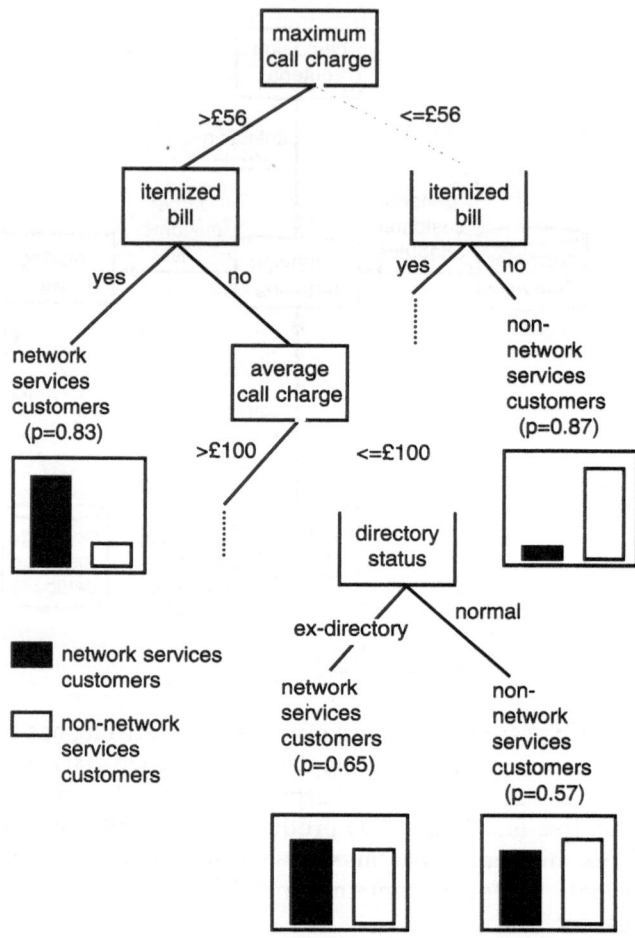

Fig. 2.8 Part of the decision tree for network services.

- by increasing response rates through better targeting, the cost of mailing is reduced;

- the generation of better targeted calling lists;

- explicit definition of the implicit market segments — giving a better understanding of the customer base and helping to design campaigns to fit the market segment;

- avoidance of annoyance on the part of those who would be uninterested.

2.2.5   Case study 4 — credit assessment

Traditional methods of credit assessment use 'score cards' [8] that are often designed and maintained by independent agencies and use data from various sources to determine credit-worthiness. Although score cards are well established they have a number of disadvantages. They are costly to develop, maintain and use, since they are owned and operated by the credit reference agency and access external bureau data. By contrast, the credit assessment system in this case study, which uses machine learning classification, relies only on the internal customer databases (see Fig. 2.9). Consequently there are few on-going costs associated with the use of the data.

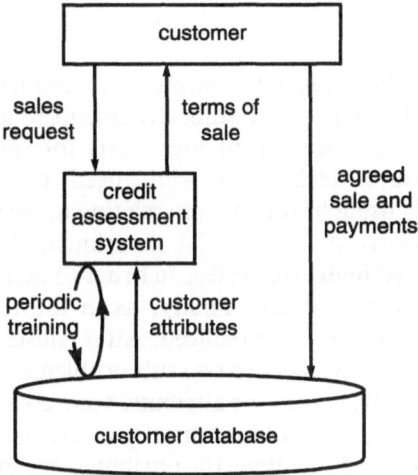

Fig. 2.9   Prediction system based on internal customer database.

Prediction models have been built using the historical data from a large sample of representative customers, using sales details, previous behaviour and actual outcomes. At the time of writing, trials involving more than 16 000 customer accounts are being monitored, with initial results showing that a high level of correct predictions have been attained.

2.2.6   Case study 5 — litigation assessment

When a customer consistently refuses to pay an invoice, few options are left, namely:

- direct customer contact — to negotiate a settlement;
- employ a debt collection agency;
- sue the customer through the courts;
- do nothing and write off the entire debt.

The customer response will be one of the following:

- to pay the full debt;
- to pay a part of the debt;
- to pay by instalments;
- to pay nothing.

This case study explored a strategy to reduce litigation costs while maintaining income, e.g. negotiating settlements to avoid defended litigation cases that would otherwise incur high costs for relatively small income.

Machine-learning techniques were used to automatically generate prediction models using historical data relating to customers' payment history and litigation behaviour. The initial experiments based on eight litigation outcomes produced high error rates. In order to achieve better classification results, two steps were taken. Firstly, as in the fault diagnosis case, the complexity of the problem was reduced. After consultation with the litigation experts, the task was divided into two sub-problems, each with different types of outcome. Secondly, composite attributes were created; the litigation experts speculated that a linear combination of two attributes could be predictive. Although some methods combine the attributes, for example neural networks, others such as decision trees do not; hence the need to create composite attributes in some cases. The resultant decision trees affirmed the expectations of the litigation experts and produced strong correlation with previously unseen historical data.

Models have been built from historical data for about 8000 customers, using previous behaviour and litigation outcomes. Predictions were made for 1300 current litigation cases. At the time of writing, the trial is still live but initial results show that a high level of correct predictions has been obtained.

2.3   CONCLUSIONS

The authors have demonstrated how volume data may be converted to high-value information through a series of case studies covering a broad range of applications. Data mining techniques have been shown to provide significant benefits either in terms of cost savings or in revenue generation. What has also been shown, particularly in the case of fraud, is that the combination of data mining with human expertise is highly effective.

A number of lessons have been learnt from these studies. Firstly, simply throwing a machine-learning system at a database is unlikely to yield good results. A significant amount of effort is required to preprocess data and understand its meaning in the problem domain. Specialist domain knowledge will almost certainly be required. Secondly, a good deal of problem simplification is likely to be needed if high accuracy results are to be obtained. This inevitably requires an element of compromise between overall business goals and what is practically achievable. Lastly, and perhaps most importantly, data mining alone will not yield business benefits. To be successful it is necessary that business processes are changed to deliver them. The mind-set which views data as something to be archived has to be changed to one which views it as a valuable resource to be exploited.

REFERENCES

1.   Frawley W J et al: 'Knowledge discovery in databases: an overview', AI Magazine (Fall 1992).

2.   Massey J and Newing R: 'Trouble in mind', Computing, pp 44-45 (12 May 1994).

3.   Integral Solutions Limited, Data Mining publicity material (October 1993).

4.   Walker G R et al: 'Visualization of telecommunications network data', BT Technol J, 11 , No 4, pp 54-63 (1993).

5.   Totton K and Limb P R: 'Experience in using neural networks for electronic diagnosis', IEE, Proceedings of Second International Conference on Artificial Neural Networks, Bournemouth, UK (18-20 November 1991).

6.   Kennett D and Totton K: 'Experience with an expert diagnostic system shell', IFIP Workshop on KBS for Test and Diagnosis, Grenoble, (27-29 September 1988).

7.   Openshaw S: 'A review of the opportunities and problems in applying neurocomputing methods to marketing applications', Journal of Targeting, Measurement and Analysis for Marketing, 1, No 2 (Autumn 1992).

8.   Lewis E M: 'An introduction to credit scoring', The Athena Press, California (1992).

3

DATA MINING — TOOLS AND TECHNIQUES

P R Limb and G J Meggs

3.1   INTRODUCTION

With the advent of powerful desktop computers, organizations are recognizing that data can be more than that. By using the appropriate tools and techniques an experienced analyst can convert voluminous data into valuable information. This can be used to highlight the success (or failure) of marketing campaigns, display processes and be more responsive to customer needs. There are a wide variety of techniques that can be employed for data analysis and increasingly the term 'data mining' is used to describe these techniques.

Data mining can be broadly categorized into three groups which:

- help build a better understanding of data;

- build characterizations of data that can be used for further analysis;

- allow the usefulness of the characterizations to be measured.

In building a better understanding of the data the analyst may, for example, use preprocessing techniques to extract features that carry the most information. Transformations may also be applied in order to reveal particular features. Two sub-groups of techniques are used to characterize data — classification and clustering. Classification techniques build models (or classifiers) from data items that are associated with a category (or class). Once derived, classifiers can be used to predict the category for data items of unknown class. Cluster analysis may give insight into the structure of large bodies of data. Measures of the usefulness of the results of different analyses can be found from statistical tests or simple visualization of results.

The variety of data mining techniques has grown, complicating the choice of an appropriate set. In addition preprocessing can have a dramatic effect on the performance of a characterization technique. Hence the permutations from raw data to classifier or cluster analysis are many, while the number of profitable routes remain few. Furthermore, techniques used for evaluating results may add complexity.

This chapter describes techniques that ease the journey from raw data to profitable characterization by providing an introduction to the broad classes of data mining techniques. The chapter begins by emphasizing the importance that should be placed on building an understanding of data prior to the application of a technique. Preprocessing techniques are then introduced with two popular techniques being discussed in more detail. Because of the sheer number of characterization techniques their discussion will be limited to that of classification (excellent introductions to clustering can be found in Anderberg [1] and Jain and Dubes [2]), with references to available algorithms being given. Finally, methods for assessing the performance of a classifier are discussed.

There is an excellent book [3] that aims to provide a comparison of classification techniques. Application areas of data mining within BT are discussed in Chapter 2.

3.2   THE IMPORTANCE OF DATA

Data mining depends heavily on the quality of the data provided. Inferences that can be drawn by computer systems from the syntactic form of data items and the semantic interpretations of their values are limited. Techniques rarely have the ability to utilize domain knowledge and certainly do not have access to the wealth of information that is at the analyst's disposal. For these reasons the analyst must ensure that the data is of the best quality possible and that as much information as possible is conveyed by the data items.

The quality of data is typified by its completeness, its separability and the extent to which it can be understood by the analyst. Data is said to be complete if records contain no missing values. The performance of a system is degraded by the presence of missing values. Several schemes have been employed to overcome this effect [4]. The separability of data refers to the ease with which classes within the dataset can be characterized. For example, a system designed to learn solutions to the travelling salesman problem would fare better if presented with cities and distances than if presented with details of the salesman's suit or contents of his suitcase. It may be a combination of attributes that carries the information necessary for distinctions to be made (see section 3.3.2). The requirement that class discriminatory features are accessible is intricately linked with the idea that the designer of the system understands the data.With)ut an understanding the designer cannot ensure such patterns are accessible.

Understanding can range from knowing data-types of attributes, to understanding statistical properties and information content. A semantic understanding of what is 'meant' by each attribute is desirable although few data mining systems are able to utilize such domain knowledge. An introduction to levels of understanding begins in the next section. The subsequent section describes how one might use visualization to build a better understanding of data.

3.2.1   Types of data

There are two fundamental data types — numeric and symbolic. Numeric data consists of both integer and floating-point numbers. Its distinguishing feature is that differences between numeric values can be interpreted as a function of differences in significance of the numbers in the problem domain, e.g. in analyzing characteristics of library visitors an attribute might be the number of books loaned in a month. A customer having loaned N books would be less significant than one having loaned $10 \times N$ books. Symbolic types may be represented by numbers or symbols, but differ from numeric types in that each possible value is of equal significance. For example, in the above scenario different titles of books would be of equal significance.

The distinction between numeric and symbolic data is necessary since data mining may treat data-types differently. Many techniques employ distance metrics to determine differences between attribute values. Such metrics make sense in the context of numeric attributes, but not in the context of symbolic attributes. Operators which test for the equivalence of values are more intelligently applied to symbolic attributes where the fact that symbols differ is of significance. If equivalence operators were applied in numeric domains the significance of differences could be overlooked. A description of data types and of reasons for distinguishing between them can be found in Anderberg [1].

3.2.2   Visualization of data

Large volume data is often difficult to interpret because it cannot easily be presented in a meaningful manner. Once a data set has more than 3 attributes or over 1000 examples it is difficult to present the data with any degree of clarity. The increase in computer use and electronic storage has encouraged the collection of data, but the corresponding development in presentation and interpretation is not so apparent (see Chapter 4). In data mining it is useful to be able to view data prior to any deep analysis, for example, to find out if there are any natural clusters in the data.

Imagine a data set in two dimensions as in Fig. 3.1.

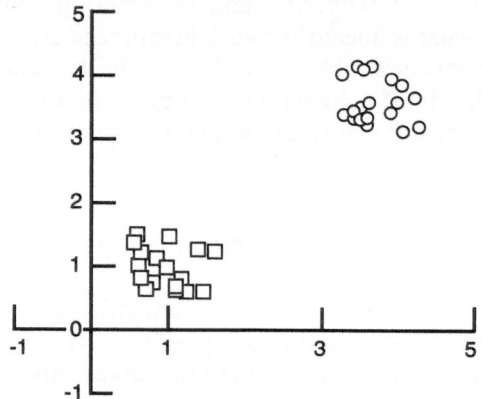

Fig. 3.1   Graph of clustered data in two dimensions.

Similar data in three dimensions can be imagined as 'clouds' of data points as shown in Fig. 3.2.

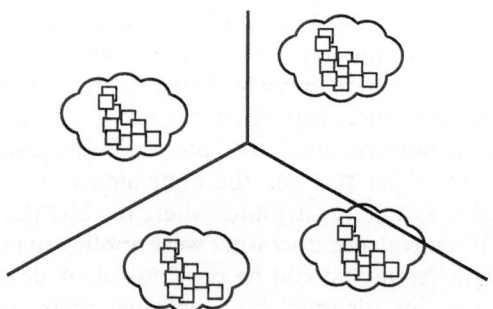

Fig. 3.2   Graph of clustered data in three dimensions.

This data might represent three attributes of a piece of equipment that fails in one of four different ways. Visualization shows that the data points which represent the four failure modes form quite distinct clusters. Data from the real world is usually not easily viewed as clusters. Two approaches to visualizing multidimensional real-world data are explained in Chapter 4.

In this section, the importance of data has been discussed and data visualization introduced. Another important step in the data mining process is preprocessing data before applying a classifier. The next section introduces some techniques for this.

3.3   PREPROCESSING TECHNIQUES

The advantages of using preprocessing techniques are twofold:

- greater insight into the data may be gained;
- many data mining techniques benefit from preprocessing rather than using the 'raw' data [5] — often this can be a simple normalization transformation.

3.3.1   Normalization of data

Theoretically, classifiers can learn to apply appropriate linear or nonlinear scaling to the input data so as to maximize classification accuracy. However, this is likely to produce classifiers that are very complex. This can be avoided by preprocessing the data. Two of the most basic preprocessing transformations that can be applied are to zero the mean and normalize the variance of the input data attributes.

Formally, zeroing the mean can be expressed as follows. If the ith attribute in the kth example of a class j pattern vector is a_{ikj}, the mean value of the ith attribute over all examples of all classes is:

$$E\{a_i\} = \frac{1}{N} \cdot \sum_{j=1}^{N} \frac{1}{\text{No. examples in class } j} \times \sum_{k=1}^{\text{No. examples in class } j} a_{ikj}$$

$$\ldots (3.1)$$

where N = number of classes and the new zero mean attributes are α_{ikj}:

$$\alpha_{ikj} = a_{ikj} - E\{a_i\} \qquad\qquad \ldots (3.2)$$

The variance of any attribute in a data set can, and often does, vary widely. A wide dynamic range can cause problems with classification techniques, particularly if a distance metric is involved. Distances between points of an attribute with large variance will dominate over other attributes, even though the attribute may not carry any more discriminatory information. It is therefore good practice to normalize the range of pattern attributes. If the ith attribute in the kth example of a class j pattern vector is a_{ikj}, the variance of the ith attribute over all examples of all classes is $V\{a_i\}$:

$$V\{a_i\} = \frac{1}{N} \cdot \sum_{j=1}^{\text{no classes}} \frac{1}{\text{No. examples in class } j} \times$$

$$\sum_{k=1}^{\text{No. examples in class } j} (a_{ikj} - E\{a_i\})^2 \qquad \dots (3)$$

where N = number of classes and the set of new attributes with normalized variance is α_{ikj}:

$$\alpha_{ikj} = \frac{a_{ikj}}{\sqrt{V\{a_i\}}} \qquad \dots (4)$$

More detail on scaling and normalization can be found in Tattersall [5].

There are many more schemes for normalizing data to enhance the accuracy or learning ability of the chosen classification technique(s). Two examples are mutual information and probability distribution normalization; these are explained elsewhere [6]. Other forms of preprocessing are used to attempt to simplify the problem by reducing the number of dimensions in the input data, thus making classifier computations less complex. There are also techniques which measure the correlation between the input values and attempt to reduce or remove the correlations such that each input describes a different dimension of the data. An example of this is principal component analysis, described in section 3.3.2.

3.3.2   Principal component analysis

Principal component analysis (PCA) attempts to reduce dimensionality by transforming correlated attributes to uncorrelated components which are a linear combination of the original dimensions.

PCA can best be described using a diagram. Figure 3.3 shows data represented in two dimensions, X_1 and X_2, the attributes of the data. Clearly there is correlation between the two attributes, i.e. as X_1 increases, X_2 increases by a corresponding amount. The PCA algorithm will transform X_1 and X_2 to new attributes, labelled X'_1 and X'_2 in Fig. 3.3, in which most of the variation is attributed to the new attribute X'_1 and the new attributes X'_1 and X'_2 are uncorrelated. The authors are indebted to Cox and Chichlowski for this explanation [7]. Detailed exposition of PCA can be found in Chatfield and Collins [8].

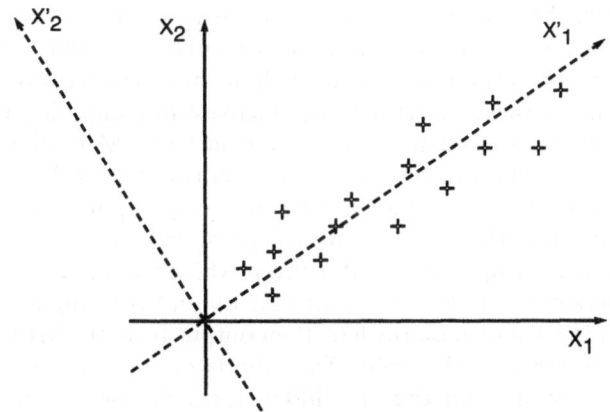

Fig. 3.3   Graph showing transformation of data using PCA.

Having introduced preprocessing of data it is now appropriate to go on to introduce a number of classification algorithms commonly used in data mining.

3.4   CLASSIFICATION ALGORITHMS

Given an understanding of the data, the choice of an appropriate algorithm can begin. Despite attempts at defining a model of which algorithms perform best on different data sets, no such model exists. In the following, no attempt is made to define which algorithms are best suited to which data sets. Instead an overview is given, the aim of which is to increase the reader's understanding of classification algorithms.

The overview describes major classes of classification algorithms giving an introduction to 'how' the learning is performed in each case. Specific instances of algorithms are referenced.

3.4.1 Classification using neural networks

Neural networks, or more correctly artificial neural networks (ANN), is a term given to a wide selection of algorithms being used increasingly in many areas of data mining. A full description of ANNs is beyond the scope of this chapter but an outline of one particular ANN is given as a starting point for further study by the reader if desired.

Research into ANNs was begun by scientists interested in how the brain functioned. A significant contribution was made by McCulloch and Pitts [9] in which they derived theorems related to models of neural systems. Recent years have seen a re-emergence of interest in ANNs. This may be attributed to the development of new algorithms which overcame some of the problems of earlier work and the increase in computing power.

A key contribution came from Rumelhart and McClelland [10] who introduced the backpropagation learning algorithm. Multilayer perceptrons (MLP) using backpropagating have become very popular in ANN research and it is this algorithm which will now be described.

MLPs are made up of layers of neurons which are simple processing units connected as a network. Figure 3.4 shows a 3-layer MLP. Input data is applied to the input layer shown on the left, then output from the MLP is produced at the output layer on the right. The middle or 'hidden' layer is used for internal processing, and thus is hidden from the user. The connections between the neurons are the means by which data are passed from one layer to the next. The connections are weighted, i.e. they have a numerical value which modifies the effect of the data passed to subsequent layers. Thus, if neuron number 1 in the input layer has a value of 0.5 and the connection between it and neuron 1 in the hidden layer has a weight of -1.1, then the value passed into neuron 1 in the hidden layer is $(0.5 \times -1.1) = -0.55$. All inputs are connected to all of the neurons in the hidden layer, thus the total value passed into neuron 1 in the hidden layer is the sum of each of the values from the input layer multiplied by their respective connection weights. More formally:

$$H_j = \sum_i w_{ji}.o_i$$

where H_j is the input to the jth neuron in the hidden layer;
$\quad\quad w_{ji}$ is the connection between the ith neuron in the input layer and the jth neuron in the hidden layer;
$\quad\quad o_i$ is the output value from the ith neuron in the input layer.

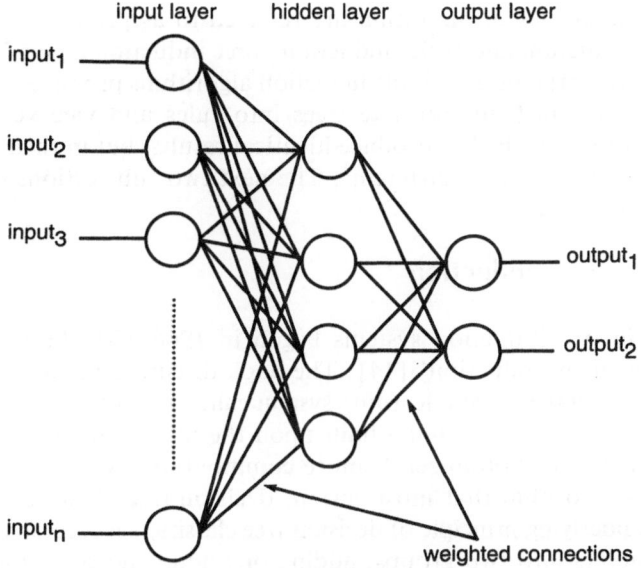

input layer   hidden layer   output layer

input$_1$

input$_2$

input$_3$

input$_n$

output$_1$

output$_2$

weighted connections

Fig. 3.4   Example structure of a neural network.

This calculation of weighted sums is also true of the output layer neurons. The MLP learns by means of the backpropagation algorithm which attempts to learn a generalized association between known inputs and known outputs in the training data thus allowing the trained MLP to make similar associations from new input data when the output is not known. The method for doing this is to minimize the error between the known output values and the output values produced by the MLP by adjusting the connection weights. This is done by propagating the error backwards through the network, hence the name. Full derivation of the mathematics behind this can be found in Rumelhart and McClelland [10].

Networks can be of unlimited size, both in terms of the number of neurons in any layer and the number of layers. In practice two layers plus an input layer are common. The number of inputs and outputs is dictated by the data and is limited by the memory and processing power of the computer running the algorithm. Further information on this and other implementations of neural networks can be found in Caudhill [11] and Lippman [12].

3.4.2   Induction of classification models

Two broad classes of algorithm are referred to as performing 'induction' — 'tree' induction and 'rule' induction. Tree induction algorithms produce decision tree structures and rule induction algorithms produce decision rules. It is a simple task to translate trees into rules and vice versa to create classification models that produce identical results, but in their purest form the algorithms are very different[1]. The next two sub-sections describe tree and rule induction.

3.4.2.1   Tree induction

Interest in tree induction systems began in 1966 with the publication of 'Experiments in Induction' [14]. The book discussed the nature of human learning and introduced a learning system called the concept learning system (CLS). In 1983 Ross Quinlan built upon the idea of the CLS and created the interactive dichotomizer 3, more commonly known as ID3 [15]. It is from this algorithm that most current decision-tree classifiers descend.

The underlying principle of decision-tree classifiers is recursive partitioning the exemplars into two groups, adding one node and two branches to the decision tree. A decision node is a test performed on an attribute, e.g. AGE $= < 65$. The process of partitioning is applied to each of the resultant groups until a stopping criteria is reached. Decision-tree classifiers differ in how decision nodes are formed and how stopping criteria are defined.

As an example consider ID3. ID3 uses information gain to form decision nodes. The information content of a group of exemplars is measured and possible partitions of that group are considered with the information gain being measured in each case. The partition resulting in the greatest information gain is chosen. Partitioning stops when no information can be gained from partitioning further, i.e. when all exemplars within a group are of the same class. An excellent example of ID3 in action can be found in Forsyth and Rada [16].

Other algorithms within the class of decision-tree classifiers include C4.5 [13], CART [17], NewId [18], Cal5 [19] and AC [20].

3.4.2.2   Rule induction

An early algorithm which uses the principle adopted by today's rule induction systems is given in Winston [21]. Winston showed that by using a small set

---

[1] In C4.5 [13], rules are generated from an examination of the tree and so the above statement is not true in this case.

of spatial and logical constructs such as 'on', 'touching', 'and', 'or' and 'not' it is possible to learn the concept of a simple structure such as an arch. Probably the most well-known example of a rule induction system is AQ-11 [22], the system which learnt to classify Soya-bean diseases more accurately than human experts.

Rule induction systems work by forming an initial set of rules based on some starting criterion. They use these rules to classify exemplars with performance being measured. Next, rules are refined by generalization, specialization and recombination to produce new rules which better classify the exemplars[2]. Again performance is measured and the process repeated until some stopping criterion is reached. Rule induction systems differ in how rules are adapted between iteration and in their stopping criterion.

As an example consider AQ-11. AQ-11 generalizes rules by adding disjunctive concepts to them. Thus a rule of the form IF [A AND B] could be generalized to the rule IF [A AND B] OR [C]. To make a rule more specific AQ-11 may add an extra conjunct to a rule. Thus a rule of the form IF [A] could be specialized by adding the conjunct AND B to form the rule IF [A AND B]. The process of generalization and specialization stops when all examples are correctly classified or when further improvement is not possible.

Other algorithms within the class of rule-induction systems include CN2 [24] and other members of the AQ family of algorithms [25].

3.4.3   Statistical classification

In section 3.2.2 it was explained that data can be considered as a set of points in multidimensional space. Classes within the data can be expected to appear as distinct clusters in the space. Parametric statistical approaches to classification use knowledge of class distributions to determine planes in the multi-dimensional space that divide classes from each other. Non-parametric statistics (considered in the next section) classify without making assumptions about the underlying distributions of classes (the authors are indebted to Taylor et al [3] for this distinction).

3.4.3.1   Parametric statistical classification

Fisher [26] first proposed the idea of a linear discriminant in 1936 to define lines in two-dimensional space (or planes in multidimensional space) that

---

[2] The process of generalization and specialization is more commonly referred to as candidate elimination. The authors feel that the explanation of this concept would introduce material that is beyond the scope of this chapter. The interested reader should refer to Thornton [23] for an introduction to candidate elimination.

separate classes. The linear discriminant uses linear combinations of attribute values in the determination of the position of the plane. Fisher uses the method of least squares which calculates both the mean-square error within groups and the mean-square error between groups, and then maximizes the ratio between them by varying the weightings on each of the attributes used. The weightings effectively define the plane which is then used for classification. Exemplars of unknown class are plotted in the multi-dimensional space and classified according to the region in which they fall. For more detail the reader is referred to Cox and Chichlowski [7].

In its basic form Fisher's linear discriminant does not use information about class probabilities. A development of the linear discriminant which does use class probabilities is linear discrimination by maximum likelihood. It has been shown to produce identical solutions to Fisher's analyses [27]. The maximum likelihood approach assigns exemplars to the class for which the value of the probability distribution function is the highest, thus effecting a decision plane where class probabilities are equal.

Variants on the linear discriminant based on the principle of maximum likelihood are the quadratic discriminant [28] which allows the surface of the plane dividing classes to be quadratic (curved) and the logistic discriminant [29] which begins by forming a plane based on the linear discriminant, but then adjusts the plane's trajectory to minimize the number of mis-classified examples in the training set.

3.4.3.2 Non-parametric statistical classification

Non-parametric statistical methods make classifications without requiring assumptions of underlying probability distributions. Methods instead make use of local density estimates of class distributions in the multidimensional space. One of the best known and most widely used methods in this class of algorithms is the K-nearest neighbour algorithm [2].

In K-nearest neighbour the position of the exemplar is conceptually plotted in the multidimensional space. The distance between it and exemplars of known class are calculated and the classes of the K-nearest exemplars are noted. The exemplar is assigned to the class that is most popular among the K-neighbours. K is usually chosen to be odd so that equal probabilities are not possible. The distance between exemplars is usually taken to be the Euclidean distance although other measures, such as the Mahalanobis distance [30], are also possible.

Other algorithms in the class of non-parametric statistical measures include density estimation [31] and project pursuit [32].

3.4.4   The use of genetic algorithms in classification

Genetic algorithms (GAs) were first introduced by John Holland [33]. Traditionally GAs were used in optimization, and search problems arrive at a near-optimal solution in a relatively short time. More recently GAs have been considered for use in data mining using three different approaches:

- firstly, in conjunction with existing classification algorithms — by finding near optimal solutions the GA can narrow the search space of possible solutions to which the traditional system is then applied, the resultant 'hybrid' [34] approach presenting a more efficient solution to problems in large domains;

- secondly, GAs have been used as post-processors of classification models [35] — neural networks, decision-trees and decision-rules are structures that can be adapted; by coding these structures it is possible to optimize their performance, e.g. in a neural network the number and organization of neurons may be adapted;

- finally, GAs may be used in the building of classifier systems [36] — genetic-based decision rules that classify in their own right; this application of GAs will be considered in more detail here.

Classifier systems, like GAs, were a product of John Holland's belief in the idea 'that simple structures undergoing simple alterations can evolve to encode tremendously complicated systems' [37]. A classifier system maintains a set of decision rules. Each rule has a strength which is a measure of performance. Examples are presented to the system and a 'competition' ensues between rules to decide which should classify the example. The competition consists of each rule making a bid which is proportional to its strength and a decision as to which rule wins. If the winning rule correctly classifies the example its strength is incremented; otherwise its strength is decremented. Rules that fail to win lose a proportion of their strength. Under certain conditions (e.g. every 'n' iterations or where performance improvements have stabilized) the GA component of the classifier system is used. The GA takes existing rules with high strength and recombines them to produce new rules which are introduced at the expense of poorly performing rules. The processes of presentation of examples, competition and GA intervention are repeated until some stopping criterion is reached. The output of the classifier system is a set of decision rules. A more complete examination of classifier systems is given in Goldberg [36].

Several distinct classifier algorithms have been introduced. The next section discusses another important factor, analysis of the results from a classifer.

3.5   POST-CLASSIFICATION ANALYSIS

Preparing data and applying classification algorithms is only part of the story. To be of any value the meaning and validity of results must be known. Usually, it will be required that results are presented to others in a meaningful manner. Again a number of techniques are available, some of which are introduced in this section.

3.5.1   Performance analysis of classifiers

For any classification method, it is necessary to know how well it is performing, and whether the algorithm is actually learning anything. Three methods are described which help to give some indication of learning performance.

3.5.1.1   Analysis using 'train and test'

A common technique is to separate the data into two parts — a training set and a testing set. Often the split is in the ratio 9:1 but others have advocated ratios of 1:1 or 4:1 [3]. The classifier is trained using the training set and then tested by applying the previously unseen test set. This test gives an indication of the error rate of the trained algorithm.

A potential problem is over-training. A classifier may be trained to such a high level of accuracy on training data that it then gives poor results on test data. The classifier has fitted or modelled the training data so well that it has lost most of its generalizing power.

A simple allegory is the least squares method of data fitting in two dimensions. Suppose a data set gives the plot shown in Fig. 3.5. It is easy to see that there is an underlying, linear trend, which can be shown using least squares (the straight line in Fig. 3.5).

If the points are fitted too closely, the result will be more as that shown in Fig. 3.6.

Interpolation of this data is equivalent to the test phase in classification. The two figures show quite different results. For the same point, at $X = 118$, Fig. 3.5 with the generalized linear fit gives a Y value of 118 while the 'over-trained' fit shown in Fig. 3.6 gives a Y value of 126.

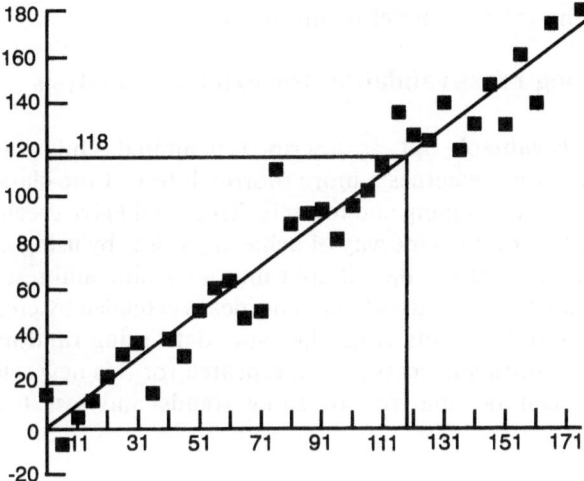

Fig. 3.5   Example of least squares fitting of data.

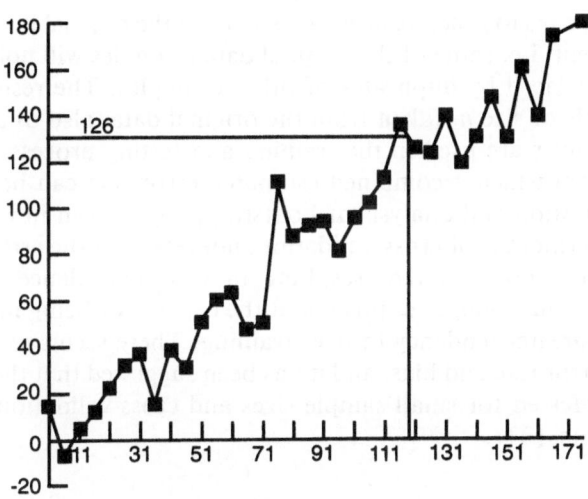

Fig. 3.6   Example of fitting data too closely.

The method of training and testing can help to give early indication that over-training is occurring. The solution is dependent on the classifier in use. A neural network can be trained to a lesser degree and/or using more training data. Tree induction can use pruning which uses a separate set of data between the training and testing stage to prune back the decision tree. In both cases, generality of the trained model is improved.

3.5.1.2 Using cross-validation for extended analysis

Train and test is valuable but, as described, is limited. Only single training and test sets are used, whereas a more thorough test of the classifier would be to use a number of training and test sets. This would give greater statistical significance to the results. One way of achieving this is by using n-fold cross-validation [38]. As before, data is split into a training and test set and the training and test processes carried out. The idea is extended by creating $(n - 1)$ further training and test sets from the same data using random selection. The training and testing process is then repeated for this new selection. The overall results can be analysed to show trends and test for statistical significance.

3.5.1.3 Bootstrap — an alternative to cross-validation

Bootstrapping aims to generate new data sets from the original by re-sampling and replacement, i.e. some of the original data examples will not be present and will be replaced by duplicates of other examples. The result is a new random sample of size n, taken from the original data, also of size n. New bootstrap samples are used in the training and testing process to produce error rates from which a combined estimated error rate can be produced. A fuller description and analysis of bootstrapping is given in Efron [39].

There are criticisms of cross-validation, not least that the error estimates it produces are too scattered, resulting in wide confidence levels. The bootstrap procedure addresses this but at the expense of being more heavily biased, i.e. a greater tendency to over training. There seems to be a trade-off between error rate and bias, and it has been suggested that the bootstrap method is preferred for small sample sizes and cross-validation for larger samples of data [3].

3.5.2 Refining classification using cost functions

Cost functions are an additional refinement to classification. Costs of misclassification are produced and then classification is based on the principle that the total cost of misclassification should be minimized. For a more detailed analysis of cost functions, see Fukunaga [40]. In practice, costs of misclassifications can be very difficult to obtain; it is hard to quantify the size of possible penalties, e.g. in a credit scoring application where costs may vary between individuals.

3.5.3 Analysing the statistical significance of classification tests

A common starting point for the statistical analysis of classification results is a confusion matrix, originally defined by Massy [41]. A confusion matrix summarizes the number of correct and incorrect classifications made by a classifier. Suppose there are two possible classifications from a set of data, A and B and the data comprises N examples. Then the confusion matrix will, in general, look like:

Actual Classification

		A	B
Predicted	A	n_{11}	n_{12}
Classification	B	n_{21}	n_{22}

where:
n_{11} is the number of examples of class A, correctly classified,
n_{12} is the number of examples of class B, incorrectly classified,
n_{21} is the number of examples of class A, incorrectly classified,
n_{22} is the number of examples of class B, correctly classified,

and $n_{11} + n_{12} + n_{21} + n_{22} = N$.

The standard statistical method for testing the significance of the results from a classifier is to use a chi-squared test [2, 42]. This test is based on the difference between observed and expected frequencies:

$$\chi^2 = \sum_1^N \frac{(O-E)^2}{E}$$

where:
N = number of data examples,
O = observed frequency,
E = expected frequency.

A large value of χ^2 indicates a large deviation from the null hypothesis, H_0, which means that the null hypothesis is not to be accepted. Since the null hypothesis is a statement that the distribution is random, a large chi-squared value indicates that the results from the classifier are not random and thus the classification is statistically significant.

A full description and derivation of chi-squared can not be given here, but any good statistical reference work will be suitable for a more in-depth study, for example Wilks [42].

3.6   CONCLUSIONS

A guide to data mining tools and techniques has been presented. The guide has deliberately been written as an overview to give the lay reader an introduction to what is a very wide ranging topic. However, the reader will note from the number and range of references that it also serves as an introduction to other works which will allow the reader to gain more detailed information in the topic(s) of interest.

Similar introductions to data mining and machine learning exist [43-45], but very often a more focused view is taken. The authors have in this chapter aimed to guide the reader through the process from the initial point of having a data set for analysis through to interpretation of the results.

The ideas presented here give an insight into the potential of data mining. A review of some application areas within BT are given in Chapter 2.

REFERENCES

1.   Anderberg M R: 'Cluster analysis for applications', Academic Press (1973).

2.   Jain A K A and Dubes R C: 'Algorithms for clustering data', Prentice Hall (1988).

3.   Taylor C, Michie D and Spiegelhalter D: 'Machine learning, neural and statistical classification', Ellis Horwood (1994).

4.   Totton K A and Limb P R: 'Electronic diagnosis using a multilayer perceptron', BT Technol J, $\underline{10}$, No 3, pp 97-102 (1992).

5.   Tattersall G D et al: 'Feature extraction & visualization of decision support data', BT Technol J, $\underline{10}$, No 3, pp 110-123 (1992).

6.   Tattersall G D: 'Investigation of data pre-processing for neural net classifiers', University of East Anglia, DOC:007, Issue 1 (Nov 1992).

7.   Cox S J and Chichlowski K: 'The application of standard statistical techniques to pattern classification', University of East Anglia, DOC:009, Issue 1 (March 1993).

8.   Chatfield M G and Collins A J: 'Introduction to multivariate analysis', Chapman & Hall (1980).

9.   McCulloch W C and Pitts W: 'A logical calculus of the ideas imminent in nervous activity', Bulletin of Mathematical Biophysics, $\underline{5}$, pp 115-133 (1943).

10. Rumelhart D E and McClelland J L: 'Parallel distributed processing, Volume 1, Foundations', The MIT Press (1987).

11. Caudhill M: 'Neural networks primer', Parts I-VIII, AI Expert (Dec 1987 to Aug 1989).

11. Lippman R P: 'An introduction to computing with neural nets', IEEE ASSP Magazine, 4-23 (Apr 1987).

13. Quinlan J R: 'Cr.5! Programs for machine learning', Morgan Kauffman, San Mateo, California (1993).

14. Hunt et al: 'Experiments in induction', Academic Press, New York (1966).

15. Quinlan J R: 'Learning efficient classification procedures and their application to chess end games' in Michalski R, Carbonnel J and Mitchell T (Eds): 'Machine learning: an artificial intelligence approach', Palo Alto! Tioga (1983).

16. Forsyth R and Rada R: 'Machine intelligence: Applications in expert systems and information retrieval', p 59-64 (1986).

17. Brieman L et al: 'Classification and regression trees', Wadsworth, Monterrey, CA (1984).

18. Boswell R A: 'Manual for NewID version 2.0', Technical Report TI/P2154/RAB54/, Turing Institute (Jan 1990).

19. Mueller W and Wysotzki F: 'Automatic construction of decision trees for classification', in Moser K and Schader M (Eds): 'Annals of Operational Research 32', J C Baltzer A G Science Publllishers, Wijdenes, The Netherlands (1994).

20. Nadel B A: 'Constraint satisfaction algorithms', Computational Intelligence, 5, Part 4, pp 188-224 (Nov 1989).

21. Winston P H: 'Learning structural descriptions from examples', in Winston P H (Ed): 'The psychology of computer vision', McGraw-Hill (1975).

22. Michalski R and Larson J: 'Incremental generation of VLI hypotheses: The underlying methodology and the description of program QA11', Urbana: University of Illinois at Urbana-Champaign, Dept of Computer Science Report (ISG 83-5) (1973).

23. Thornton C J: 'Techniques in computational learning', Chapman & Hall (1992).

24. Clark P and Niblett T: 'Induction in noisy domains', in Brakto I and Lavrac N (Eds): 'Progress i machine learning', Sigma Press (1987).

25. Michaelski R, Mozetic I, Hong J and Lavrac N: 'The multi-purpose incremental learning system AQ15 and its testing application to three medical domains', in Proc AAAAI-86, California, Morgan Kauffman (1986).

26. Fisher R A: 'The use of multiple measurements in taxonomic problems', Annals of Eugenics, 7, pp 179-177 (1936).

27. Michalski R, Mozetic I, Hong J and Lavrac N: 'The multi-purpose incremental learning system AQ15 and its testing application to three medical domains', Proc AAI-86, California, Morgan Kauffman (1986).

28. Clarke W R et al: 'How non-normality affects the quadratic discriminant function', Comm Statistics — Theory and Methods IT-16, pp 41-46 (1979).

29. Cox D R: 'Some procedures associated with the logistic qualitative response curve' in Dvid F N (Ed): 'Research papers on statistics: Festschrift for J. Neyman', pp 57-77, John Wiley, New York (1966).

30. Mahalanobis P C: 'Historical note on the D3-statistic', Sankhya 9, p 237 (1948).

31. Fix E and Hodges J L: 'Discriminatory analysis, nonparametric estimation: consistency properties' Report 4, Project 21-49-004, USAF School of Aviation Medicine, Randolph Field, Texas (1951).

32. Freidman J H: 'SMART's user guide' Technical Report No 1, Laboratory of Computational Statistics, Department of Statistics, Stanford University (1984).

33. Holland J H: 'Adaptation in natural and artifical systems', Ann Arbor: The University of Michigan Press (1975).

34. Kelly J and Davis L: 'Hybridizing the GA and K-nearest neighbors classification algorithm', in Proceedings of the Fourth International Conference on Genetic Algorithms, California (Jul 1991).

35. Dodd N: 'Optimization of network structure using genetic techniques', AIENG-91: Applications of artificial intelligence in engineering 6: Proceedings of the sixth international conference. Oxford UK, pp 939-944 (July 1991).

36. Goldberg D: 'Genetic algorithms in search, optimization and machine learning', (1989).

37. Davis L: 'Genetic algorithms and simulated annealing', Pitman, London (1987).

38. Stone M: 'Cross-validatory choice and assessment of statistical predictions', J Roy Statist Soc, 36, pp 111-33 (1974).

39. Efron B: 'Estimating the error rate of a prediction rule: improvements on cross-validation', J Amer Stat Ass, 78, pp 316-331 (1983).

40. Fukunaga K: 'Introduction to tatistical pattern recognition', Academic Press (1972).

41. Massy W F: 'On methods: discriminant analysis of audience characteristics', J of Advertising Research, 5, pp 39-48 (1965).

42. Wilks S S: 'Mathematical statistics', John Wiley & Sons (1963).

43. Weiss S M and Julikowski C A: 'Computer systems that learn', Morgan Kauffman (1991).

44. Gordon A D: 'Classification', Chapman & Hall (1981).

45. Piatesky-Shapiro G and Frawley W J: 'Knowledge Discovery in Databases', AAAI Press (1991).

4

VISUALIZATION TECHNIQUES FOR DATA MINING

G D Tattersall and P R Limb

4.1 INTRODUCTION

Data mining is a term which has become popular to describe a number of techniques for the exploration and exploitation of data. In particular, a large part of data mining involves the visualization of data and subsequent utilization of machine-learning techniques for data classification. This chapter describes some techniques for data visualization which enable the user to enhance understanding of the structure and properties of data. Such insight into the nature of a data set is very useful when deciding what type of preprocessing should be applied prior to automatic classification [1, 2], or prior to application of machine-learning techniques for further analysis and exploration (see Chapter 3).

BT collects and stores large quantities of data from a variety of sources. These large data sets typically describe different states of a system and are difficult to interpret because there is no obvious way of abstracting and presenting data features in a meaningful way for a human observer. Examples of such data range from credit status of customers of a company, to acoustic data generated from speech. The use of computers has made it relatively easy to collect and store such data, but this has not been accompanied by a corresponding development in methods of interpreting and displaying the data.

A particular entry in a database typically consists of the values of a number of measurements by which the data is described. For example, a single entry in a database relating to customers might provide values for the number of lines rented by the customer, the last bill value and the customer's title. The number of lines, bill value and title are called attributes and may have numerical or symbolic values such as '£347' or 'Mr'. It is common practice to imagine individual data examples, described by N attribute values, as points in an N-dimensional space. The co-ordinates of the point in the space are the attribute values. This way of imagining data leads to the idea of visualizing the distribution and relationship between data examples by 'taking a walk through the N-space'. In general, many more than three attributes are used to describe a data example and it is difficult to conceive such a dimensional space.

Visualization of the data examples in a high-dimensional space can be made much easier by forming a two-dimensional map of the N-dimensional space. This is not a new idea and an algorithm developed by Sammon [3] in 1967 has been widely used. However, this algorithm is severely limited when applied to large data sets, because its computational complexity rises as the square of the number of data points to be mapped. Other methods of visualizing data structures have been proposed in Batchelor [4] which project the pattern points from the N-space on to a pair of reference vectors.

The work described here is a novel approach to mapping which has been developed by the authors. This approach, which is based upon a neural network, produces a map very rapidly compared to other more conventional techniques, and its speed of operation allows the user to interact with the data while looking at the map. For example, the distribution of the data in the map can be observed as the user interactively modifies various parameters, such as the way in which the distance between data examples is measured in the high-dimensional space. Interactive mapping enables the user to discover natural clusters of examples in a data set, and also find features which can be used in automatic classification of the data.

The chapter also shows how different types of two-dimensional data maps can be generated to emphasize different aspects of the data. For example, it is possible to generate maps which simply reflect the point-to-point distances between data examples in their N-space, or which emphasise the separation between clusters of data points in the N-space. Perhaps most importantly, a map can be generated to show the degree of separability of the data points in terms of their assigned class. The latter map can be used to assess how well the data could be automatically classified, as well as highlighting those specific examples which are liable to be misclassified.

All of these interactive data-mapping facilities, which have been made possible by the new algorithm, have been embodied in a data visualization system called the hidden target mapping (HTM) tool which enables the user to interact with the data map displayed on the screen via the computer's mouse. This tool is being further developed to make it part of a suite of data mining tools.

Before describing the HTM interactive visualization tool in detail, the operation of the Sammon mapping algorithm is reviewed in order to highlight some of the problems of mapping N-dimensional data on to a two-dimensional map.

4.2 THE SAMMON MAP

4.2.1 The Sammon mapping algorithm

Successful interaction with the data visualization tool requires fast mapping of multi-dimensional data on to a two-dimensional map for display with as little distortion as possible of the apparent distances between examples in the data set. A well-established approach to such mapping was first proposed by Sammon [3], but unfortunately the practical use of the Sammon mapping algorithm is limited because its computational complexity rises as the square of the number of data points to be mapped. Typically, if the algorithm is run on a PC, a map of 20 data points can be generated in a few seconds, but would take hours for a thousand data points. The long time to generate a map makes it impossible to use the algorithm in an interactive manner and therefore effectively mine the data.

However, the Sammon algorithm exhibits many of the important aspects of data mapping and is described in detail here to illustrate general problems and show the state-of-the-art prior to the development of the mapping tool.

The basis of the Sammon mapping algorithm technique is to map the set of data points in N-dimensional space. The relative positions of the two-dimensional points are iteratively modified until their relative distances mirror as closely as possible the relative distances between pairs of points in the N-dimensional space.

The iterative adjustment of the positions of the points in two-dimensional space is done using gradient descent minimization of a mapping error, E, which is defined as the average of the squared difference between the distance of each pair of points in the N-dimensional space and the corresponding pair of points in the two-dimensional space.

Let the ith N-dimensional vector in the set of patterns to be visualized be denoted as X_i and the corresponding two-dimensional vector be Y_i. Assuming that there are a total of M examples in the data set, there are M points in the N-dimensional space and two-dimensional space. The squared distances between the ith and jth vectors in the N- and two-dimensional spaces are $d_p(X_i, X_j)$ and $d_m(Y_i, Y_j)$ respectively where $x_{i,k}$ and $y_{i,k}$ are the values along the kth dimension of X_i and Y_i respectively:

$$d_p(X_i, X_j) = 2 \sqrt{\sum_{k=1}^{N} (x_{ik} - x_{jk})^2} \qquad \dots (4.1)$$

$$d_m(Y_i, Y_j) = 2 \sqrt{\sum_{k=1}^{2} (y_{ik} - y_{jk})^2} \qquad \dots (4.2)$$

A measure of the total difference between the distances of pairs of points in the N-dimensional and two-dimensional spaces is therefore:

$$E = \sum_{i=1}^{M} \sum_{j=1}^{M} \left\{ d_p(X_i, X_j) - d_m(Y_i, Y_j) \right\}^2 \qquad \dots (4.3)$$

Usually the initial values of Y_k are set randomly and are iteratively modified using the steepest descent algorithm defined in equation (4.4) in which k_s is a small constant which determines the learning rate. The gradient term is evaluated by differentiation of equation (4.3):

$$y_{ik}^{n+1} = y_{ik}^{n} - k_s . \frac{\partial E}{\partial y_{ik}^{n}} \qquad \dots (4.4)$$

4.2.2   Problems of the Sammon mapping algorithm

There are four particular areas of the Sammon mapping algorithm that cause concern.

- Computational complexity — any mapping algorithm must work sufficiently fast that a user can make changes interactively to the data while observing the map. This means that the mapping must converge within a fraction of a second after any changes have been made. The computational complexity of the standard Sammon algorithm makes this impossible unless the number of examples being displayed in the map is very small.

- Placing new data in the map — if a new data example is to be added to an existing Sammon map, the entire data set with the new data example needs to be remapped. It is difficult to simply place the new example in an existing map.

- Map storage — maps of data need to be stored for future reference in exactly the same way as normal geographical maps. The Sammon map requires storage of the co-ordinates of every single point in the map and the map cannot be represented in a compact form.

- Topological order — a map is said to be topologically ordered if the position ordering of points in the map reflects their position ordering in the original space. This kind of topological order can be global, in which case the ordering of all points is correct, or local, in which case only the ordering of points which are close together is correct.

The standard Sammon map attempts to reflect global topological order and local topological order of the space in which the data examples are described. This leads to a conflict in the mapping unless the data examples actually lie on a plane in their space. Experiments have indicated that the Sammon map generally minimizes the conflict by effectively projecting the data points on to a suitably oriented plane rather than the surface of a complex manifold. Projection on to a plane is probably a useful solution, but it can be achieved much more simply by techniques other than Sammon mapping.

4.3   THE HIDDEN TARGET MLP MAPPING ALGORITHM

4.3.1   Basis of the new mapping algorithm

The new mapping algorithm described in this chapter is based on the multilayer perceptron (MLP) [5]. The MLP is iteratively trained using a modified form of error backpropagation called the hidden target mapping algorithm. This algorithm effectively projects the examples of N-dimensional data on to a curved surface or plane within the N-dimensional space, and then displays the 'flattened out' surface with the projected data points as a two-dimensional map. The surface is automatically positioned to maximize the accuracy of the map and the amount by which the surface curves is controlled by selecting an appropriate MLP architecture.

4.3.2   How the algorithm solves the computation, new data, and map storage problems

The key to the algorithm's reduced computational load is that it places sensible constraints on the form of the mapping generated and, in exchange, requires much less computation than other techniques such as the Sammon map. The degree of mapping constraint is determined by the complexity of the MLP and can range from forcing all points to lie on a plane, to placing points on a highly complex curved surface. In the former case very little computation is required to form a map, while in the latter considerable computation is needed. In comparison, the Sammon mapping algorithm places no constraints on its mapping and as a result requires still more computation to form a map.

The algorithm also solves some of the other problems raised in connection with the Sammon mapping algorithm — in particular the problems of new data and map storage. If the HTM algorithm has already been used to generate a data map and it is desired to place a new data example in the existing map, the new data example is simply applied as a pattern to the input of the MLP. The output of the MLP will be the co-ordinates of the position of the new example in the existing map.

The HTM algorithm also solves the potential problem of map storage associated with the Sammon mapping. The entire HTM map can be regenerated from the weight values of the MLP used to form the mapping. A typical MLP used in this application might use 250 connection weights, each represented by a four-byte floating-point number. Thus only 1 kbyte of memory is required to store the map, regardless of the number of points in the database which are to be mapped.

4.3.3   Structure of the HTM MLP

The MLP used to form the map has N input units to which the N-dimensional data example is applied, a number of hidden units, and two output units whose output values are the co-ordinates to which the input is mapped in the two-dimensional map. The output units are linear and the hidden units use sigmoidal activation functions.

The function of the MLP is to develop a linear or non-linear transform of the input N-dimensional pattern to an output two-dimensional pattern which can be plotted as a point in a two-dimensional map. The form of the transform is determined by the weight values of the MLP which are iteratively adjusted to minimize an error measure for the mapping. This is shown in the next section.

4.3.4   The HTM training algorithm

It is usual to train an MLP using error backpropagation [6] in which the derivative of the error between the MLP's outputs and specified target values are evaluated and used to adjust the weights of the MLP using the least mean squares (LMS) algorithm. However, in the mapping application, no explicit output target values are available, and a different form of error must be evaluated.

The mapping error measure chosen for the HTM algorithm is essentially the same as that used in the Sammon mapping. The distance, $d_p(X_i, X_j)$, between each pair of data examples, X_i and X_j, and the N-dimensional pattern space is compared with the distance, $d_m(Y_i, Y_j)$, between the corresponding pair of points Y_i and Y_j in the map space. The weights of the MLP are iteratively adjusted to minimize the square of the difference between $d_p(X_i, X_j)$ and $d_m(Y_i, Y_j)$ over all the examples in the data set.

For practical reasons, the MLP weights are updated after the evaluation of the mapping error for a randomly selected pair of data examples, instead of after the mean square mapping error for the entire data set has been evaluated. This is equivalent to 'per example training' as opposed to 'epoch training' of MLPs. The algorithm for the iterative modification of the weights therefore proceeds as follows:

- initialize all the weights in the MLP with small random values;

- randomly select two N-dimensional pattern examples, X_i and X_j, from the data set;

- record the responses of the MLP, Y_i and Y_j to the inputs X_i and X_j;

- evaluate the distances $d_p(X_i, X_j)$, and $d_m(Y_i, Y_j)$;

- evaluate the squared error, $e^2(i, j)$, between $d_p(X_i, X_j)$, and $d_m(Y_i, Y_j)$ and then evaluate the derivatives of the squared error with respect to each weight in the MLP;

- use the gradient descent algorithm to update each weight value;

- go to the second stage and repeat the process.

4.3.5   Evaluating the error derivatives

The squared error derivatives used in the gradient descent adaptation of the weights of the MLP are evaluated by differentiation of the expression for the squared error between the N-space distance, $d_p(X_i X_j)$, and the distance between the corresponding two-dimensional vector outputs from the MLP, $d_m(Y_i, Y_j)$, with respect to the output from the MLP, i.e:

$$e^2(i,j) = \left(d_m(Y_i, Y_j) - d_p(X_i, X_j) \right)^2 \qquad \text{... (4.5)}$$

and

$$\frac{\partial e^2(i,j)}{\partial y_{ik}} = \frac{2\{d_p(X_i, X_j) - d_m(Y_i, Y_j)\} \cdot (y_{ik} - y_{jk})}{d_m(Y_i, Y_j)} \qquad \text{... (4.6)}$$

It is then simple to evaluate the required derivatives of the squared error with respect to the weight values of the MLP by using the derivative given in equation (4.7) in conjunction with the standard backpropagation rule [5] in which the term $\partial_{1,k}$ is set equal to the derivative of the squared error with respect to the kth output of the MLP:

$$\delta_{1k} \; = \; \frac{\partial e^2(i,j)}{\partial y_{ik}} \qquad \qquad \ldots (4.7)$$

The derivative of the squared error with respect to a weight, $w_{n,k,s}$, which connects the output of the kth neuron in the $(n+1)$th layer to the sth unit in the nth layer, is then given by the recursively evaluating expressions (4.8) and (4.9) in which S_n is the number of neurons in layer n, and the output of the kth neuron in layer n in response to input X_i, is denoted by $o_{i,n,k}$:

$$\delta_{n+1,k} \; = \; o_{i,n+1,k} \, (1 - o_{i,n+1,k}) \, . \, \sum_{s=1}^{S_n} w_{n,k,s} . \delta_{n,s} \qquad \ldots (4.8)$$

$$\frac{\partial e^2(i,j)}{\partial w_{n,k,s}} \; = \; o_{i,n+1,k} . \delta_{n,s} \qquad \qquad \ldots (4.9)$$

The weights of the MLP can then be updated iteratively using gradient descent, as shown in equation (4.10), in which p is the iteration index:

$$\omega_{n,k,s}^{p+1} \; = \; \omega_{n,k,s}^{p} \; - \; k_s \, . \, \frac{\partial e^2(i,j)}{\partial \omega_{n,k,s}} \qquad \qquad \ldots (4.10)$$

4.3.6   Forcing local topological order

The error measure (equation (4.5)), which is minimized by the HTM algorithm, makes no distinction between the mapping error for pairs of points which are distant in the pattern space and pairs of points which are close together. This is precisely the problem mentioned in connection with the standard Sammon map. Inevitably the mapping will attempt to simultaneously reflect the global and local topological order of the space in which the data examples are described, and this may lead to a conflict in the mapping unless the data examples actually lie on a plane.

The problem is ameliorated by the presence of the term $d_m(Y_i, Y_j)$ in the denominator of the expression for the error derivative given in equation (4.6). If two points are close together in the map, $d_m(Y_i, Y_j)$ is very small and tends to amplify the value of the error derivative. Thus the mapping becomes much more sensitive to errors in mapping points which are close together rather than far apart, and the map will tend to reflect local topological order at the expense of global order. This property has been exploited in the HTM algorithm by including a locality control which can be modified while the map is being displayed. The locality control is a numerical factor which is used to control the influence of the $d_m(Y_i, Y_j)$ term in the denominator of equation (4.6). The locality control, k_L, is introduced into the modified expression for the error derivative given below. The range of k_L is zero to one. When it has zero value, the denominator of equation (4.11) becomes independent of $d_m(Y_i, Y_j)$ and global order will tend to be reflected in the mapping. When k_L is one, the $d_m(Y_i, Y_j)$ term will have full effect and local order will tend to be reflected in the mapping at the expense of global order:.

$$\frac{\partial e^2(i,j)}{\partial y_{i,k}} = \frac{2\left\{d_p(X_i, X_j) - d_m(Y_i, Y_j)\right\}.(y_{i,k} - y_{j,k})}{d_m(Y_i, Y_j).k_L + 1 - k_L} \qquad \text{... (4.11)}$$

4.3.7   Practical considerations in the use of the HTM algorithm

4.3.7.1   Absolute values of outputs

An MLP trained using the HTM algorithm is only provided with an error related to the relative positions of the points in the map space, and not their absolute values. This can lead the MLP to produce a set of points, $\{Y\}$, which have massive absolute value even though their relative values are correct. This may result in numerical overflow. A solution is to modify the expression (4.7) for δ_{ik} so that it includes a term which causes the values of the outputs to tend towards zero. This is achieved by including a weighted absolute target value for $y_{i,k}$ of zero, as shown in equation (4.12), in which the tendency for the outputs to adapt to zero value is controlled by the constant k. Typically k has a value of 0.05:

$$\delta_{1k} = \delta_{1k} - y_{ik} \qquad \text{... (4.12)}$$

4.3.7.2   Choice of single versus multilayer HTM algorithm

It is frequently found that the N-dimensional data can be mapped quite accurately on to a plane rather than a curved surface. To do this, the MLP is made so that it has just a single layer of linear units. This has a number of beneficial effects — the speed of learning is high, convergence to a global minimum is certain, and a map is immediately visible even before weight adaptation has started. The latter property arises because any set of weights in a linear perceptron will define a surface on to which the data can be projected.

Using a single-layer MLP will not always allow a satisfactory map to be generated. If it is found that the normalized mean square mapping error does not converge to a small value when using a single layer MLP, it will be necessary to introduce a second layer of units. A practical approach is to start by using a small number of hidden units in the second layer. If this fails to provide a satisfactorily low mean-square mapping error, the number of hidden units should be increased.

4.4   PATTERN-SPACE DISTANCE METRICS AND THE HTM SYSTEM

4.4.1   The effect of changing the distance metric

HTM is iteratively modified until the relative values of the two-dimensional output of the HTM's MLP mirror as closely as possible the relative distances between pairs of points in the N-dimensional space. It makes sense to plot points in the map space using simple Euclidean distance (equation (4.1)) because humans intuitively understand this type of distance. However, there is no reason why some other metric should not be used in the N-dimensional pattern space. If this is done, the Euclidean distances in the map space become proportional to distances in the N-space according to the other chosen metric.

There are numerous pattern-space distance metrics which could be used in conjunction with the mapping, and some examples which have been built into the prototype visualization tool are presented in the following section.

4.4.2   Examples of pattern-space distance metrics

4.4.2.1   Vari-power metric

This metric is a generalization of the Euclidean distance metric. The distance between two vectors is found from the summation of the distances along each dimension, raised to the power p, which can be any real positive value. If p is made large, then the distance between two vectors tends to be dominated by the largest distance along any particular dimension. Conversely, using a value of p which is much less than one tends to make the distances measured along each dimension have equal significance, regardless of their actual value. The metric is defined as follows:

$$D(X_i,X_j) = p \sqrt{\frac{1}{N} \sum_{k=1}^{N} |x_{ik} - x_{jk}|^p} \qquad \ldots (4.13)$$

4.4.2.2   City block metric

The city block metric is a special case of the vari-power metric in which the power, p, is equal to one. When applied to binary data, this metric effectively returns the Hamming distance. The metric is defined as follows:

$$D(X_i,X_j) = \frac{1}{N} \sum_{k=1}^{N} |x_{ik} - x_{jk}| \qquad \ldots (14)$$

4.4.2.3   Normalized dot product metric

The normalized dot product metric returns a similarity, $S(X_i,X_j)$, value which is proportional to the cosine of the angle between a pair of vectors, independent of their magnitudes:

$$S(X_i,X_j) = \frac{1}{|X_i|} \cdot \frac{1}{|X_j|} \sum_{k=1}^{N} x_{ik} \cdot x_{jk} \qquad \ldots (4.15)$$

However, for the purposes of generating maps, a distance is required rather than a similarity value, and the following conversion, shown in equation (4.16), has been chosen which returns a distance value in the range 0—1:

$$D(X_i,X_j) = \frac{1}{2} (1 - S(X_i,X_j)) \qquad \ldots (4.16)$$

4.4.3   Attribute weighting in the map

In addition to selecting a particular type of pattern-space distance metric, it is possible to weight the importance of each attribute such that it contributes more or less to the calculated pattern-space distances. The weighting of each attribute can be changed interactively while the map is being displayed, and this facility enables the user to examine the significance of each attribute.

Typically, the examination would be done by setting all attribute weightings to zero except the weighting of the attribute being tested. If the resulting map shows class-specific clusters, it is clear that the particular attribute is important. Conversely, if increasing the weight of a particular attribute does not cause greater class or cluster separation in the map, it indicates that the attribute is not useful.

The attribute weighting can also be used to control the viewing aspect of the map. This happens because the HTM error is minimized by primarily projecting the values of the weighted attributes on to the map. Thus, changing the attribute weighting function changes the 'angle' of the data projection on to the map surface.

4.5   PRODUCT FEATURES AND THE HTM ALGORITHM

Different categories of data are often defined by the co-occurrence of pairs of attributes. One approach to numerically indicating the co-occurrence of a pair of attributes is by forming a new feature which is the product of the values of the chosen pair. If the product has a positive value, it indicates that the attributes simultaneously have the same polarity and vice versa. (The product must be evaluated after the chosen attributes have been normalized by setting their mean values to zero.) The product features from each pattern in the data set can be evaluated and then displayed in a map using the HTM algorithm. This may reveal structure in the data which is not apparent when the original attributes are mapped.

An interesting and potentially useful case arises when a product feature is formed by squaring a single attribute value. Regions in the original attribute space which are enclosed by a convex boundary will be mapped to a region which can be separated by a single hyperplane in the feature space.

The arithmetic process of forming product features can be applied equally well to both real-valued and binary-symbolic data. In the latter case it is equivalent to forming a new feature which is the 'exclusive-OR' of the two chosen attributes. This may or may not be useful, but it suggests that future work should allow new features to be generated interactively, which are any user-defined logical function of the original binary-symbolic attributes.

4.6   USING HTM TO EMPHASIZE DIFFERENT
ASPECTS OF THE DATA

4.6.1   Class distance based maps

The HTM algorithm attempts to find a linear or nonlinear projection of the N-dimensional patterns on to a two-dimensional surface which is displayed as the data map. The projection is adjusted until the point-to-point distances in the N-space are reflected as nearly as possible in the two-dimensional map space.

This idea can be extended to force the projection to reflect some externally defined point-to-point distances in the N-space. In particular, if the user arbitrarily defines the distances between each pair of points in the N-space, the HTM algorithm can use these distance values to adjust the projection until they are reflected as nearly as possible on the map. The only change to the HTM algorithm is that the user-defined point-to-point distances are substituted for the pattern space point-to-point distances.

The inter-class distances are defined in the form of a matrix of values which are typically set to unity. This causes the HTM to attempt to place all examples of the same class at the same position in the map and make all classes equidistant. Using a two-dimensional map, this is only possible if the number of classes is less than or equal to three. In practice it has been found that using a two-dimensional map in conjunction with four or more classes still works satisfactorily, although the mapping error increases as the number of specified classes increases.

4.6.2   Natural cluster distance-based maps

The aim of this type of mapping is to cause the N-dimensional data to be projected into the two-dimensional map in such a way that natural clusters in the data are displayed with maximum separation. The chosen approach is to perform a cluster analysis on the data in its N-space. The N-space distances between the centroids of each cluster are then measured and the HTM is adjusted so that the map space reflects the inter-centroid distances as nearly as possible. The entire data set is then subjected to this mapping and displayed on the map.

There are several possible clustering algorithms which could be used in this application, but the most suitable candidate appears to be a simplified form of the algorithm used in the Kohonen network [6]. The basis of the clustering is as follows.

- Before the mapping commences, assign random positions to the centroids of the k clusters into which the data is to be grouped. Note that the value of k can be adjusted interactively whilst the program is running. It would be usual to start with a low value then progressively increase the value of k, all the while observing to see if widely separated clusters are being generated.

- Randomly select a pair of data examples from the data set. This is the same pair as used in each iteration of the HTM algorithm. For the purposes of clustering, only one example is needed. However, since two are selected for the HTM algorithm, it is computationally efficient to use both at each iteration of the clustering algorithm.

- Measure the distances from both the selected examples to each of the current data centroids.

- Assign each of the two selected examples to the clusters with the nearest centroids.

- Update the centroids of the selected clusters, such that the centroid of each cluster moves fractionally closer to the corresponding data example, that is:

$$G^i = (1 - \alpha).G^i + \alpha.X^i \qquad \qquad \ldots (4.17)$$

where G^i is the centroid of the ith cluster to which the selected data example X^i has been found to belong, and α is a small constant which determines the rate at which the centroid is updated.

- Repeat the process starting at the second stage.

4.7   EXAMPLES OF THE USE OF THE HTM TOOL

A simple example which illustrates the basic operation of the HTM tool is shown in Fig. 4.1, which shows the map formed by HTM of the points on lines of 'longitude' and 'latitude' on a sphere. Points in the northern hemisphere are labelled as class 1 and points in the south, class 2. It is evident that there is no way of unambiguously mapping a three-dimensional distribution into a two-dimensional space and so the map is simply a projection of the sphere on to a plane.

The type of mapping used in this example is based upon the pattern-space distances.

A more interesting example is shown in Figs. 4.2-4.4 which show HTM maps for articifial nine-dimensional data which contain nine Gaussian distributed clusters at each vertex of a nine-dimensional hypercube. Three different classes have been arbitrarily defined, with three clusters being associated with each class.

class 1+ class 2 ×

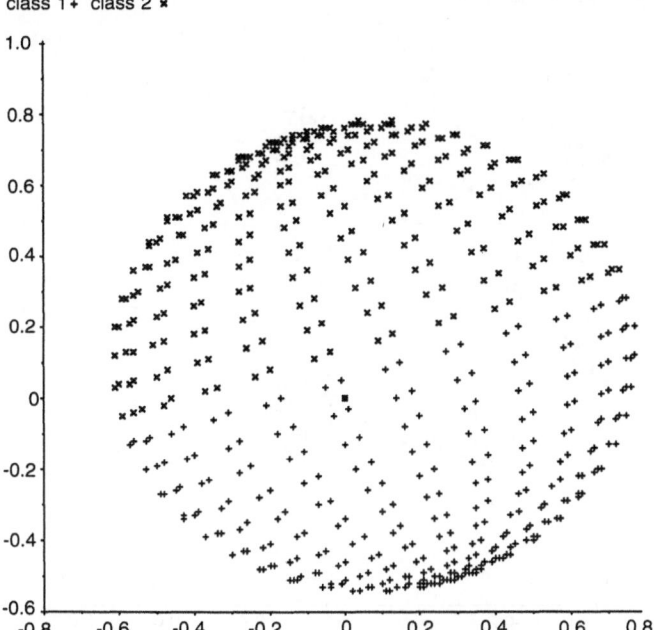

Fig. 4.1   HTM map of points on a sphere.

class 1• class 2 + class 3 ×

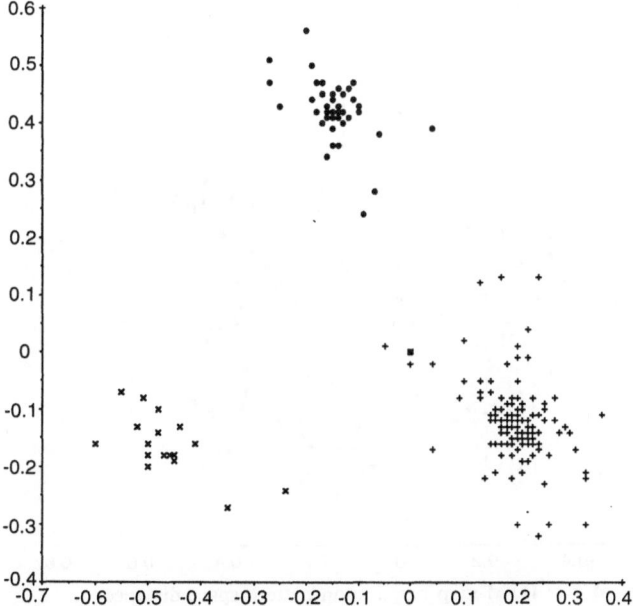

Fig. 4.2   Class-distance HTM map of nine-dimensional data.

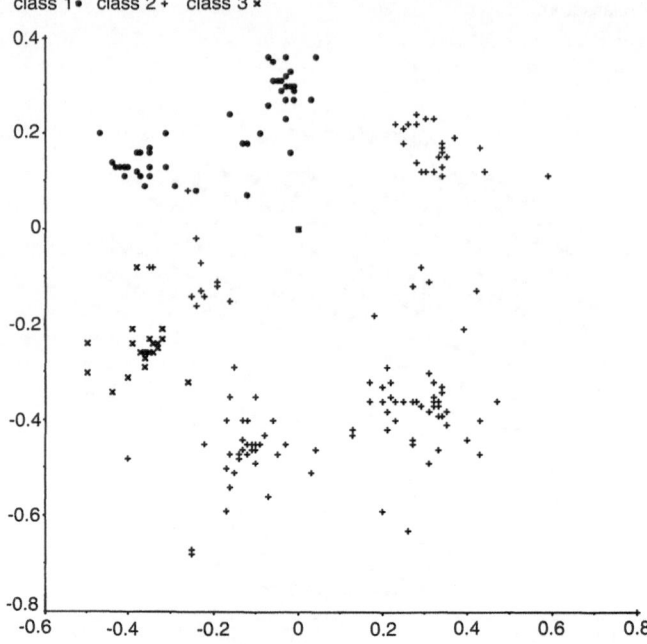

Fig. 4.3   HTM based upon natural-cluster distances.

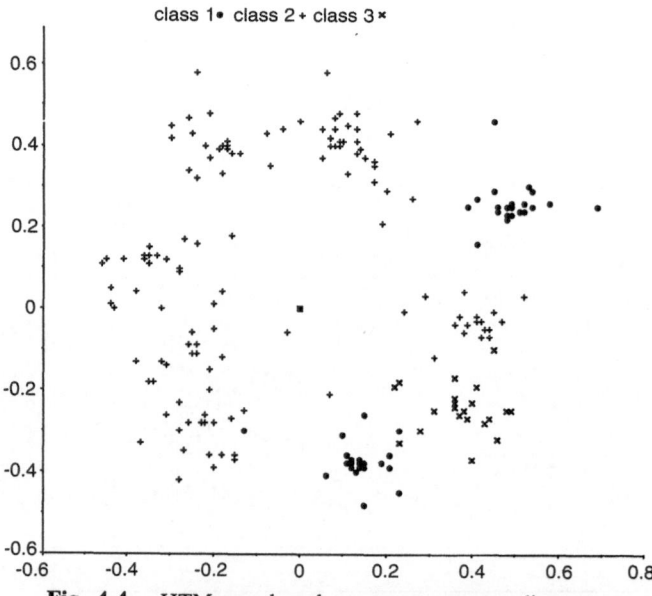

Fig. 4.4   HTM map based upon pattern-space distances.

The map in Fig. 4.2 is based upon class distances, i.e. the map has been generated to maximize the distance between examples of different classes. The map in Fig. 4.3 has been generated to maximize the distances between natural clusters in the data, and the map in Fig. 4.4 has been generated to reflect the point-to-point pattern-space distances in the nine-dimensional space.

Finally, maps of some real seven-dimensional financial data are presented in Figs. 4.5 and 4.6. Each example in this data set has been categorized into one of two classes.

It can be seen from the pattern-space distance-based map in Fig. 4.5 that the classes generally occupy different class regions, and that there is some clustering into sub-groups. (Investigation of the sub-groups revealed that they were caused by one of the attributes in the data only taking integer values.) It can be seen from the class distance-based map in Fig. 4.6 that the classes would be rather easily separated using a single discriminant plane, although several examples are positioned ambiguously and would be misclassified using such a classifier.

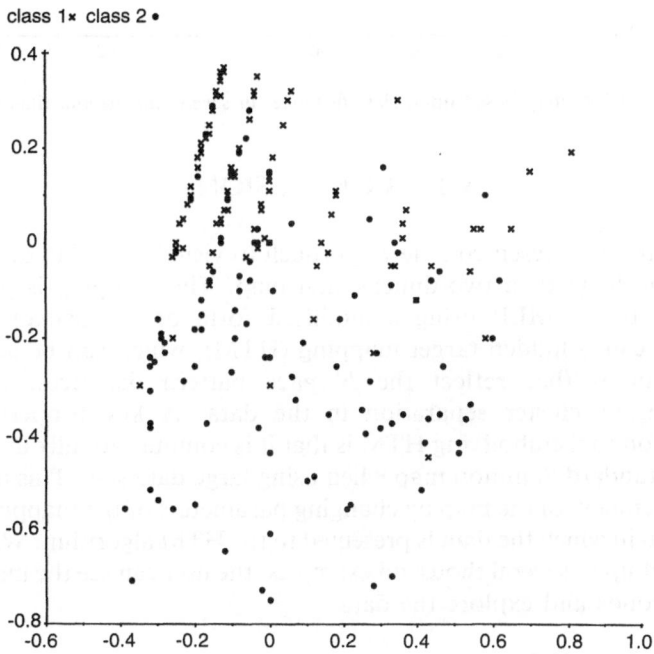

Fig. 4.5   HTM map based upon pattern-space distances in seven-dimensional financial data.

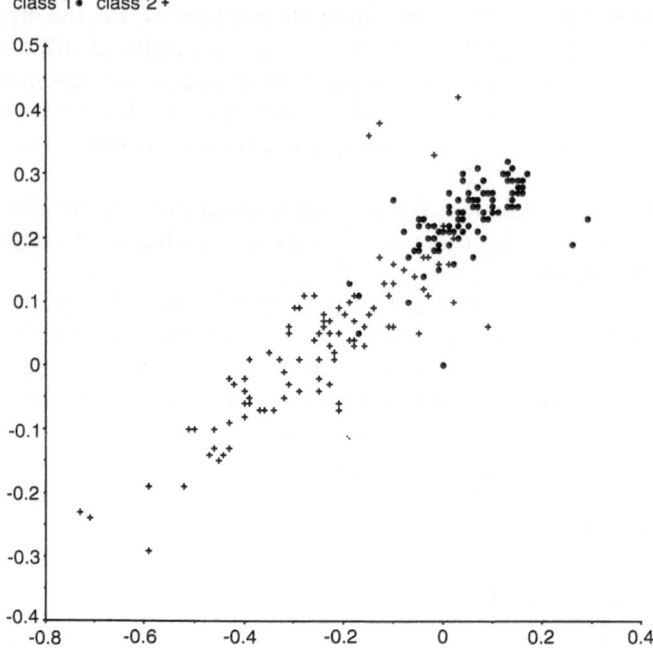

Fig. 4.6   HTM map based upon class distances in seven-dimensional financial data.

4.8   CONCLUSIONS

This chapter has presented a new approach to visualizing high-dimensional data in the form of a two-dimensional map. The mapping is performed iteratively by an MLP using a modified form of ·the backpropagation algorithm called hidden target mapping (HTM), which can be adapted to generate maps that reflect the N-space pattern distances, inter-class separation, or cluster separation in the data. A key property of the visualization tool embodying HTM is that it is computationally much faster than the standard Sammon map when using large data sets. This allows the user to interact with the map by changing parameters of the mapping as well as the form in which the data is presented to the HTM algorithm. When using data sets of up to several thousand examples, the user can see the map change within seconds and explore the data.

REFERENCES

1. Tattersall G D, Chichlowski K and Limb R: 'Pre-processing and visualization of decision support data for enhanced machine classification', Intelligent Systems Engineering Conf, Edinburgh (August 1992).

2. Tattersall G D, Chichlowski K, Limb R and Totton K A E: 'Feature extraction and visualization of decision support data', BT Technol J, $\underline{10}$, No 3, pp 110-123 (July 1992).

3. Sammon J W: 'A nonlinear mapping for data structure analysis', IEEE Trans on Computers, $\underline{C-18}$, No 5 (May 1969).

4. Batchelor B G (Ed): 'Pattern recognition, ideas in practice', Plenum Press (1978).

5. Rumelhart D E, Hinton G E and Williams R J: 'Learning internal representations by error propagation', in Rumelhart D E , McLelland J L and the PDP Research Group (Eds): 'Parallel distributed processing', MIT Press (1978).

6. Kohonen T: 'Clustering, taxonomy and topological maps of patterns', Proc Int Conf on Pattern Recognition (October 1982).

5

NEW DIRECTIONS IN INVESTMENT APPRAISAL

N J Davies and I Napier

5.1 INTRODUCTION

In a highly competitive business environment, the effective exploitation of modern investment appraisal methods is essential if organizations are to maintain or improve their competitive position. This chapter will discuss a number of modern investment appraisal tools and techniques which have been developed or are under investigation within BT.

Whole-life costing is a technique for quantifying the long-term financial implications of a purchasing or technology selection decision so that the most cost-effective choice can be made. A wide range of potential costs (and, where appropriate, revenues) are taken into account, rather than just purchase costs, which, taken alone, are often misleading. This broader picture can lead to significant cost savings in the longer term.

Whole-life costing, and the financial analysis techniques it incorporates, are described. Consideration is given to the place of whole-life costing and appraisal in large organizations, particularly with respect to customer/supplier relationships and the procurement process. WINA, a PC-based whole-life costing tool, and its development is then discussed. Finally, some possible further developments of whole-life costing and WINA are considered, in particular, potential applications in the areas of marketing and the environment.

Other investment appraisal activities within BT are then discussed, and ITIA, a computer-based investment appraisal tool for IT investments, is briefly discussed. The technology road-map, a tool for providing a picture of the different parts of an investment programme and their interrelationships, is also described.

5.2   A WHOLE-LIFE APPROACH TO INVESTMENT APPRAISAL

5.2.1   Whole-life costing

In common with other organizations, the need to reduce costs is widely recognized within BT. To achieve real reductions, it is necessary that a medium to long term view is taken, particularly when taking decisions which affect the specification, design and selection of products and services of all kinds which are to be owned or used for a significant period.

The danger of considering only up-front costs is often illustrated by the iceberg analogy (see Fig. 5.1). Clearly, it is impossible to choose the best of a range of available options or strategies without a thorough understanding of the likely consequences of each, in terms both of costs and of benefits. The key to good decision making is reliable information — the whole-life approach described here provides this by the use of a standardized process and computer tools applicable to a range of business decisions, including supplier and technology selection.

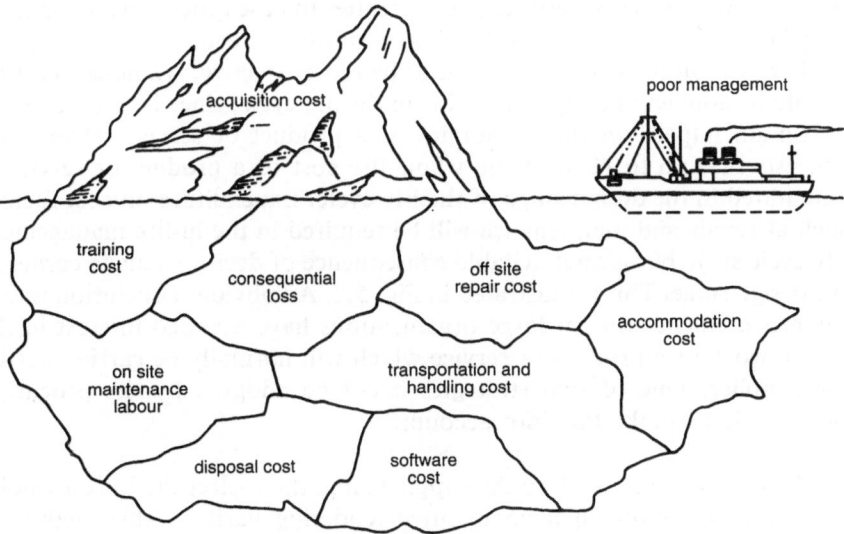

Fig. 5.1   The whole-life costing iceberg.

As discussed by Paul [1], the technique of whole-life costing was initially applied to the procurement of military equipment by the US Department of Defense in the late 1960s. This followed the realization that post-procurement costs were far greater than purchase costs over the lifetime of the equipment. Gibbs [2] states that it was estimated that only 26% of the $126 billion 1979 US Defense budget was spent on the procurement of new products, compared with 61% which was spent on operation and support costs. Care must be taken, however, not to assume that all these costs could be scaled down if equipment were more reliable or maintainable. As explained by Proffitt [3], many of these operating costs arise from the need to plan for active battle service, not just routine peace time maintenance.

Attempts were made during the 1970s to apply whole-life costing more widely. In the UK, the DTI established a National Terotechnology Centre in 1975 with the objective of promoting the adoption of whole-life costing techniques to physical asset management in British industry. However, even as late as the 1980s, as reported in Woodward [4], there was little evidence that UK managers had modified their view that, when acquiring new capital assets, initial cost was the most important consideration.

In the late 1980s, following the realization of the importance of the technique in investigating the impact of hardware reliability on costs, a whole-life costing unit was set up within BT. Whole-life costing was seen as a management tool which could be used to support decision making. Its primary objective was to identify and quantify the whole-life cost of competing suppliers' products and services and thereafter to base procurement decisions on whole-life rather than (misleading) acquisition costs.

It was soon recognised [5] that to be most effective, whole-life costing should be applied as early as possible in the life cycle, since it can then have maximum impact on the economics of a product or service. Drury [6] suggests that up to 80% of the whole-life cost of a product or service is committed in the design stage of the life cycle. Expenditure on cost factors such as repair and maintenance will be required in the in-life management life cycle stage but is an inevitable consequence of decisions made earlier in the design stage. This is illustrated in Fig. 5.2. An obvious conclusion is that the procurement arms of large organizations have a vested interest in the design work of a product or service which will normally be carried out by the supplier. One of two strategies could be adopted by the procuring organization to take this into account:

- leave the design work to the suppliers and then select the lowest whole-life cost option on a competitive tendering basis — this method is currently in use by many large organizations and involves a formal

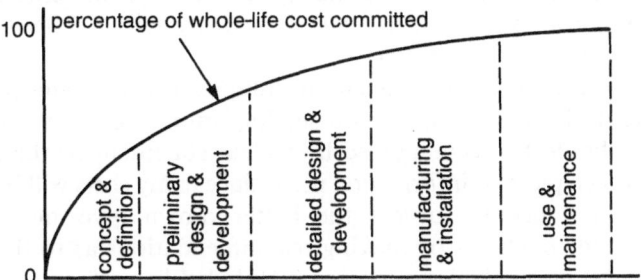

Fig. 5.2   Costs committed in life cycle stages.

transfer of technical information from supplier to procurer in tender documents to allow procurement analysts to assess the cost implications;

- work more closely with suppliers so that they have a fuller understanding of their customers' cost drivers and can influence their design work accordingly to reduce them — this involves a more iterative, two-way transfer of information.

The second approach is in principle the best, although it requires a high degree of trust between customer and supplier. Suppliers need to be certain that the customer will consistently use whole-life costing as the financial criterion at tender adjudication, so that their efforts to reduce whole-life costs will not result in lost business because the customer reverts to a decision based on purchase price. Conversely, the customer needs to be assured that the supplier cannot exploit the relationship to increase purchase prices without producing the desired reductions in ownership costs and that all commercially sensitive information passed to the supplier will remain secure.

5.2.2   Financial analysis in whole-life costing

In order to decide which of a range of options had the lowest whole-life cost, the cost factors pertaining to each option (purchase, maintenance, training, disposal, and so on) could be estimated and then summed to provide a figure for the whole-life cost. However, this would ignore the fact that the different cost factors are typically incurred at different times — purchase costs will be incurred on acquiring the product, maintenance costs will be incurred annually from installation until disposal, and so on. To address this, allowance must be made for the time value of money — a preference for spending money at some future date as opposed to spending it now. Future expenditure is preferred because money in the bank has interest-earning potential. For example, if £100 is required next year only £93 need be invested this year (assuming 8% interest).

There are a number of financial techniques for assessing the value of investment decisions. It is, of course, important that any use of these techniques should be accompanied by a clear statement of the confidence attached to each piece of information used, since many costs will be estimates.

This section briefly surveys the better-known investment appraisal techniques and concludes by making recommendations as to the preferred technique. Detailed calculations are not included here, simply a discussion of the different techniques and their strengths and weaknesses.

In financial terms, the objective of a profit-making organization is to create wealth by producing products and services with greater value than the resources consumed.

To create wealth now, the present value of all anticipated income must exceed the present value of all anticipated expenditure. All the techniques described below aim to quantify the wealth created by a given project and are collectively termed return on investment (ROI) techniques [7].

5.2.2.1   Payback period

Probably the simplest method of investment evaluation, the payback period technique yields a ratio which expresses the length of time it takes for a project to return to an organization the expenditure involved in the orignal investment. It indicates the time taken for corporate funds to be replenished by the income generated from the activity supported by the investment.

The payback method is inadequate for several reasons. Firstly, it involves the subjective establishment of an acceptable payback period (often cited as 2.5 years). Furthermore, it may ignore big paybacks later than the break-even point and it does not take into account the time value of money.

The discounted payback period technique is a modification of the payback period method, and deals with a criticism of it, in that the payback period is calculated only after discounting the cash flows (income and expenditure). Discounted cash flows (DCFs) reduce (i.e. discount) the cash flows of future years to an amount which represents their worth at the present time. The investor is then able to compare the cost at today's prices of the proposed expenditure with the evaluation (also in present day terms) of the anticipated future cash receipts.

Discounted payback thus improves on the payback method by taking into account the time value of money but is still open to the other criticisms made of it.

5.2.2.2   Internal rate of return

The internal rate of return (IRR), also referred to as the yield method, determines the rate of interest which, when applied to future cash flows will reduce (discount) their monetary value until they are equal to the cost of the investment. IRR thus calculates a percentage rating which can be used by a financial analyst to assess a project — it tells what interest rate of return is necessary for the project to break even. The rate is thus used as an indicator of investment worth — typically, the analyst would accept proposals where IRR is above a certain minimum rate.

5.2.2.3   Net present value

This technique calculates the monetary difference between the income and expenditure of a project when both are discounted at a specific percentage rate, i.e. the net present value (NPV) of a project is the sum of all cash flows (inflow and outflow) discounted at the required rate of return. The higher the NPV, the better the return on the project.

Net present value can be used to take into account the risk associated with a project — the higher the perceived risk, the higher the discount rate applied. This means that higher revenues are required to achieve a positive NPV.

The NPV and IRR techniques always give the same results when assessing a project on an 'accept or reject' basis — projects with positive NPV must by definition have IRRs greater than the required rate of return. However, while the IRR method demonstrates a percentage rate of growth, NPV reveals an absolute measure of growth in monetary terms. Furthermore, where a portfolio of possible investments is being considered, the sum of the NPVs of the investments chosen will indicate the total increase in wealth arising from that selection. No such results can be obtained from a summation of IRRs since project size is in no way reflected in IRR.

A further disadvantage of the IRR technique is that in some situations, multiple solutions are possible for solving the equation used to calculate IRR. While some authors suggest ways to overcome this problem and also assert that the conditions giving rise to it are uncommon, the fact remains that such instances can and do occur.

For these and other reasons, most commentators recommend the use of the NPV approach to investment appraisal.

In whole-life costing terms, if a number of options are being considered for possible purchase, the option with the highest (least negative) NPV should be chosen. Of course, if revenues as well as costs are being modelled, then the NPV may be positive. The whole-life investment appraisal tool described in the next section can model revenue as well as cost.

The number of years over which a costing is to be performed must relate to the economic life of the product or service. At the end of the costing period, a residual value can be estimated, based on the disposal value of the item and an estimate of any revenue that may be earned beyond the costing period.

5.2.3   Computer-based tools for whole-life costing

The key stages in all whole-life costing studies are twofold.

- Estimates are produced for annual costs of different cost factors such as purchase, maintenance, power consumption, installation, and so on. These can be arrived at using hard historical data to calculate the costs, or by using established 'rules of thumb', expert engineering judgement or sophisticated models. Typically, someone with detailed knowledge of the system under consideration would make these estimates, using a combination of these methods. Estimates of revenue may also be made.

- These estimates are then processed using financial analysis techniques taking into account inflation, the time value of money and the tax implications. As discussed above, the output would normally be a net present value (NPV) for the option under consideration.

In the 1980s, the available computer tools to support whole-life costing were identified. Five packages were evaluated, of which only CASA (cost analysis strategy assessment) from the US Department of Defense was considered adequate to examine costs across the whole product life cycle.

5.2.3.1   WINA functionality

WINA enables the user to perform a whole-life costing of a set of options, taking into account costs incurred at all stages of the life cycle of a commodity and using a BT standard methodology. Users can perform complex whole-life analyses in a Windows™ environment without specific programming or spreadsheet knowledge.

The WINA tool divides a commodity's life cycle into six stages. At each stage, the user may select a range of cost factors which contribute to the whole-life cost of the commodity. Some examples from each of the six stages are:

- product definition — market research, feasibility;

- design and development — research, tooling, test and evaluation;

- pre-launch planning — training, documentation;

- product launch — delivery, quality assurance, installation;

- in-life management — maintenance, power, accommodation;

- product withdrawal — disposal, upgrade.

Cost factors may also be automatically selected by the user from templates which provide a standard list of cost factors for different types of commodity (different cost factors would, for example, be applicable when costing personal computers from those applicable when costing company cars).

Each cost factor may be provided by the user as an annual 'lump sum', or calculated from data about the product provided by the user, for example, to calculate maintenance costs, the reliability of the commodity and the labour rate of the maintainer would be key cost drivers.

Furthermore, the user may specify how a commodity is composed of its component parts, e.g. a network switch might be composed of a power supply, a number of linecards, and so on. This component hierarchy can be built up by the user in a graphical display on the PC screen using the latest PC windowing technology. The user can then provide data at the component level; costs are then calculated at this level and 'rolled up' to give an overall system cost. Figure 5.3 shows a typical WINA screen, including a simple system component hierarchy.

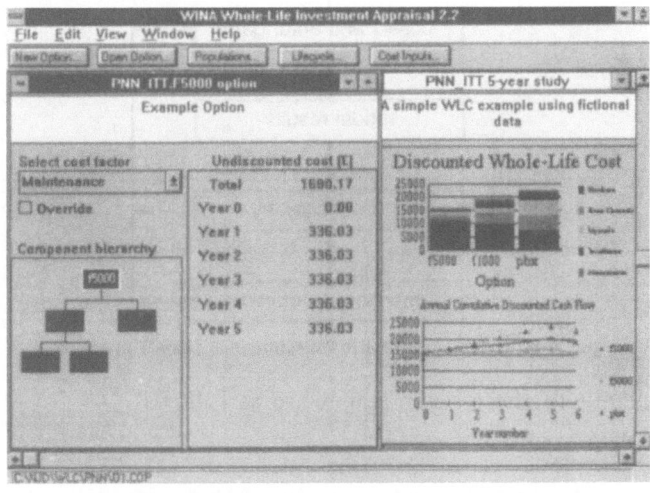

Fig. 5.3   WINA screen.

The user also provides population profile information about the commodity, which defines the time-scales and volumes in which it will be purchased, delivered, installed and withdrawn.

The costs are calculated taking into account inflation, the time value of money and corporation tax to give a net present value costing of the commodity over the chosen study period. This financial analysis enables the time at which costs are incurred to be taken into account and allows a more meaningful comparison of the options under consideration.

The resultant costings can be displayed in a variety of graphical formats, facilitating the comparison of competing options. 'What-if' type calculations are supported, allowing the user to see the result of changing the parameters of the study on the whole-life costs. Tabular outputs are also provided and both formats of results can easily be copied to other Windows applications.

A simple, seven-stage process for whole-life investment appraisal is shown in Fig. 5.4. The WINA system supports this process by providing a modelling structure and templates where appropriate.

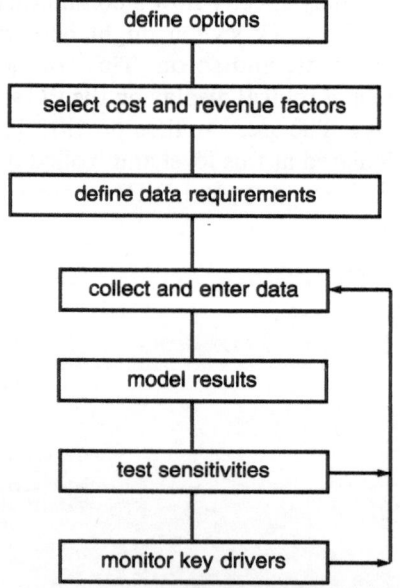

Fig. 5.4   Whole-life investment appraisal process.

The details of these seven stages are as follows.

● Define options

The first stage is to define the options under consideration. This involves setting the system boundaries, which can range from selecting a supplier for a simple product to comparing two fundamental network options. In addition, quantities and purchase profiles need to be defined (e.g. 20 000 fax machines per month for 2 years, 5 transmission systems per year, etc). As mentioned above, WINA provides facilities for population profile entry and a powerful system hierarchy builder. Any number of options can be compared.

- Select cost and revenue factors

The factors selected will depend partly on the mode in which the tool is being used. For visibility of total costs, all relevant factors should be used. For decision support, factors should be selected which are:

— likely to be of significant financial value;

— likely to have different values for different options;

— not already committed (sunk costs).

Cost/revenue factors are selected from a comprehensive menu in WINA, which can also store templates to model the life cycle of specific product types. WINA divides a product's life cycle into six stages; the user may select at each stage from a range of cost factors which contribute to the whole life cost/revenue.

- Define data requirements

When cost/revenue factors are entered, WINA prompts the user for the data required to model each. For example, if maintenance costs are to be calculated, WINA requires information on the mean time between failures (MTBF) of the units making up the system.

- Collect and input data

Input data should be gathered from equipment manufacturers, existing field databases and expert opinion as appropriate.

- Model results/output

WINA produces overall results in tabular and graphical form (Fig. 5.5), after processing the input data.

- Test sensitivities

WINA allows 'what-if' tests to be easily performed. Inputs can be varied and new graphical output produced on screen within seconds. In this way, the 'key drivers' can be identified.

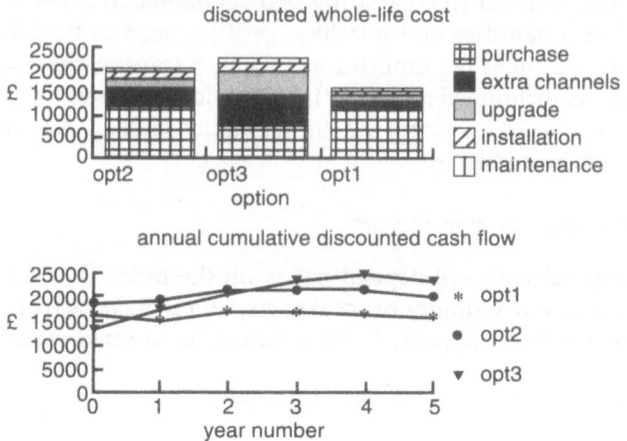

Fig. 5.5   Graphical results from the WINA system.

- Monitor key drivers

  Key drivers identified by WINA should be monitored during the lifetime of the equipment if possible. Real results can then be produced to check against the original prediction.

5.2.3.2   WINA implementation

Implementation employed a rapid prototyping approach to software development, whereby an initial prototype of the software is developed quickly and relatively early in the software development process. This prototype is shown to the potential users of the system being developed and their comments are incorporated in the next prototype of the system. This loop is iterated several times until a finished system is obtained which meets the users' requirements.

The programming language used was Smalltalk, an object oriented language available on PCs under the Windows operating system for which the system was to be developed. It is not the intention here to discuss in detail the advantages (and disadvantages) of adopting an object oriented language for software development. A relatively brief discussion is given here; the interested reader should refer to Cusack and Cordingley [8] for further information.

The fundamental organizing principle in object oriented programming languages is the combination of both data and procedures into structures (called objects) related by some form of inheritance mechanism. The conventional notion of a program as a series of steps to be carried out sequentially is replaced by a set of objects each of which has a set of operations defined over it. Instead of representing such objects as passive data structures, object oriented systems permit objects a more active role, in which they are capable of passing messages to other objects, asking them to perform certain tasks. This property makes object orientation particularly suited to simulation and applications which require detailed representations of real-world objects and dynamic relationships between them.

It was mentioned earlier that objects are related by some form of inheritance. One class of object (e.g. cat) can be defined as a subclass of another (e.g. animal), its superclass. Cat will then inherit all the data and operations of animal. In this way, a new object (cat) can be defined simply by the way it differs from an existing related object (animal). This is, in general, more efficient than defining a new object from scratch. In this way, object oriented programming encourages the reuse of code — only those operations wherein the subclass differs from the superclass need be rewritten.

A further benefit of object oriented programming is the use of abstraction; an object is essentially an abstraction of some object in the real world which it is wished to model in the program and it is only necessary to describe those aspects of the object relevant to the application at hand, with irrelevant detail suppressed.

These properties are particularly useful when using a rapid prototyping approach to software development. This is because the modularization of code and data, encouraged by object oriented programming and its reuse of code, facilitate the rapid modification of a prototype based on user feedback.

The use of inheritance also aided development, as the following (somewhat simplified) example shows. In WINA, a superclass CostFactor was created with an operation GetCosts which asks the user for the costs pertaining to a particular cost factor for each year over which the whole-life costing study was being performed. This operation simply presents users with a table into which they can enter the relevant costs.

For each cost factor (maintenance, power, training, and so on) a subclass of CostFactor was created. For some cost factors, the costs are modelled using a set of arithmetic equations. Maintenance costs, for example, can be calculated from the MTBF (mean time between failure), the MTTR (mean time to repair) and a number of other pieces of data. For these cost factors, a new GetCosts operation is created, attached to the relevant CostFactor subclass (MaintenanceCostFactor, for example) which prompts the user for the relevant data and then performs the appropriate calculation. All the other cost factors are left unchanged and inherit the simpler GetCosts operation from its superclass, CostFactor.

Smalltalk was chosen over and above the more 'conventional' object oriented programming language C++ for several reasons, not least because object oriented programming is fundamental to Smalltalk and the whole programming environment is built around it, rather than being derived from procedural programming. Smalltalk provides the flexibility of an interpreted language with powerful tracing and stepping facilities in which to test code, and the ability to modify code and variable values 'on the fly', coupled with a compiler for improved efficiency when system development is completed. These reasons, along with the existence of a powerful GUI (graphical user interface) building tool made Smalltalk the ideal choice for rapid prototyping development.

Another possible implementation vehicle for WINA was a spreadsheet system. However, WINA has a potentially wide user base, with a correspondingly wide range of computer literacy.

The interface provided was designed with the advice of human factors experts and is considered to be more friendly and intuitive, requiring less computer familiarity than could have been achieved with a spreadsheet solution. Furthermore, some parts of the WINA system such as the system component hierarchy builder (Fig. 5.3) are much less natural to implement in a spreadsheet than in a language like Smalltalk.

5.2.4   Typical applications of whole-life costing

5.2.4.1   Pair gain systems

A comparison of an analogue system with a more modern digital system revealed that the analogue system was cheap to buy and install, but, because of the lack of diagnostic facilities, it was difficult and time consuming to identify which part of the system was responsible when a failure occurred.

Conversely, the digital system was expensive to buy but the system was more reliable and had much better fault-diagnostic facilities. Using whole-life costing, it was possible to demonstrate that over a 15-year life, significant savings would be made if the more expensive digital system were purchased.

5.2.4.2   Desktop computers

Of the competing suppliers, only one was able to provide the following services with the purchase of their hardware:

• preload software at manufacturing stage;

• deliver to end users;

- provide initial training.

This supplier was shown to have considerable scope for cost savings and improved delivery timescales.

5.2.4.3   Large line concentrators

This study revealed that savings could be made by switching from the existing supplier. Before the tender, there were two suppliers — one for large line concentrators and one for small line concentrators. The offering from the supplier of the small line concentrator offered significant savings in training costs through integration of two courses and reduced spares costs through the use of common equipment.

5.2.5   Whole-life costing — the future

5.2.5.1   Further development of WINA

An obvious area for further development would be in the area of sensitivity analysis. The system already provides 'what-if' facilities to investigate quickly the result of changing key cost drivers on the whole-life cost of a system.

A further enhancement to the system would be for WINA to identify the key parameters itself; varying the values of some items of data may have little or no effect on the bottom line whole-life cost. An automatic facility to determine the most important cost drivers, and present the user with an analysis of the results of allowing them to vary, would be a useful addition to the system.

User experience may identify cost factors which are currently not modelled within the system and these could be added to the 60 cost factors already contained in WINA.

5.2.5.2   Extending the use of whole-life costing

As we have already seen, products and services which find a market primarily on the basis of their acquisition costs may not be those which incur least cost in the longer term. Those organizations offering premium quality products and services would benefit from the use of whole-life costing as a marketing technique — typically such products have a higher initial cost than those of competitors, but offer advantages over them. In many instances, these advantages can be quantified in monetary terms in a whole-life costing study. For example, improved reliability will reduce the cost of repair and maintenance during the in-life management phase of the product life cycle. Similarly, a more efficient machine will lead to lower energy costs in the longer term.

The use of whole-life costing techniques can clearly identify these longer term financial benefits of products which seem more expensive when only purchase price is considered. As such, it is a potentially powerful marketing tool in the hands of the premium quality product or service provider. An interesting side-effect of the use of such a tool in a marketing environment is that the tool itself can become a market differentiator, establishing the supplier as more technically competent in the eyes of the customer.

In an increasingly competitive market-place, the promotion of the use of telecommunications services for environmental reasons can lead to a growth in the overall market. Furthermore, an early awareness of the possibilities of this provides telecommunications organizations with a competitive edge over their rivals by allowing them to take maximum advantage of any resultant market growth whilst promoting environmental improvement. Association with protection of the environment can also enhance the image of an organization.

As Oakley [9] points out, telecommunications services can be viewed as components in a transport and communications infrastructure which facilitates the movement of goods, people and information. These three areas have all grown rapidly this century. A significant opportunity for environmental improvement is via the substitution of telecommunications for transport services. Obvious examples would be the substitution of teleworking for commuting or videoconferencing for meetings where all participants are physically present. Briefly, increased use of telecommunications could mean:

- reduced travel;
- reduced construction;
- improved information;
- more effective use of time;

which in turn can lead to:

- lower carbon dioxide emissions;
- reduced consumption of natural resources;
- lower pollution levels.

Whole-life costing has a major role to play for managers who want to investigate these possibilities. Whole-life costing can be used to consider the competing costs of a telecommunications option against its transport or other alternatives (e.g. teleworking versus commuting, videoconference versus physical meeting). A whole range of cost factors can be modelled, including those relevant to the environment, to give the true cost of each of the competing alternatives. Furthermore, the costs that are specifically environmental can be factored out to show that, in environmental terms, the telecommunications option is superior. Of course, a whole-life costing analysis would be purely financial and could be further enhanced by consideration of less tangible benefits accruing from the use of more environmentally friendly telecommunications services.

5.3   RELATED WORK

5.3.1   Appraisal of IT investment

Recognition of the potential impact of IT systems on the strategic position of companies and increasing levels of IT spend have made the control and justification of IT investment a critically important issue [10]. At the same time there is widespread doubt concerning the suitability of traditional methods of investment appraisal for the evaluation of IT proposals. The major reason for this is the intangible nature of many of the benefits to be gained by the deployment of IT systems [11].

5.3.2   Future directions

Central to the emerging technology investment appraisal techniques is the idea that projects and programmes are best viewed as strategic competitive weapons. There are many pressing reasons why there should be investment in technological programmes to ensure maximum strategic impact, including:

- compelling evidence of a strong positive relationship between the proportion of turnover that an organization invests in new technological development and its long-term profitability (see, for example, Old [12]);
- the increasingly competitive business environment, in which the rapid and sustained delivery of high-quality, cost-effective new products and services to meet customer requirements is essential;

- the need to balance investment in support of and to expand existing areas of business with the need to undertake longer term research and development to provide new products and services in the medium and long term.

The above considerations have led to the recognition of the importance of clear links between technological investment and corporate strategy [13]. This requires a shift away from discrete, project-based investments towards a more coherent, integrated investment in technology where the business needs and strategy of the organization are given the highest priority. Business managers as well as technical managers are actively involved in the control of the consequent portfolio of technical activities which will best fulfil the goals of the organization as a whole.

Technology road-maps are increasingly being used [14] to provide a co-ordinated view of the various components of an overall portfolio or programme of technical activities of the type previously described. Individual projects can be seen as part of the whole programme and monitoring of progress against overall business objectives is possible.

A road-map is essentially a diagram constructed from facets, threads and ties. Facets are major work areas within a programme and appear diagrammatically as horizontal bands on the road-map. The particular facets required on a road-map will depend on the nature of the programme. Consider a programme to investigate and develop some new network service — the facets might be:

- enabling R&D;
- technology integration investigation;
- concept demonstrator development;
- field trial;
- skill development;
- training;
- roll-out.

For each facet, there may be a number of threads. For example, the 'enabling R&D' facet could have threads for the identification of relevant technology, university contract(s), technology transfer, in-house research, and so on. Threads are represented as labelled horizontal lines within the appropriate facet. Ties show major connections between programme elements. They are shown on road-maps as diagonal or vertical lines showing the supporting relationship between the elements. Ties are one way in which lower level technical activities can be presented in terms of higher level business objectives. A (somewhat simplified) part of the example road-map discussed above is shown in Fig. 5.6, showing the ties 'improve technical awareness' and 'prove technical feasibility'.

Current areas of interest are the use of road-map overlays to provide different perspectives on a programme, the representation of risk, and how progress could be related to more traditional financial investment appraisal.

Road-maps have benefits for investment appraisal and management, including:

- placing projects in context and justifying them in terms of their contribution to broader strategy;

- improved communication between users, technologists and managers;

- more effective communication of programme progress to all participants and managers;

- identification of omissions that need to be addressed.

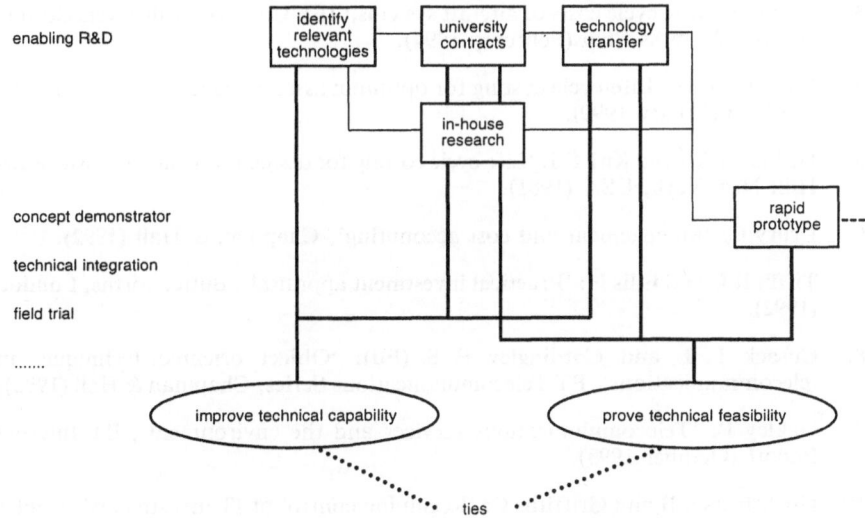

Fig. 5.6 Extract from a technology road-map.

5.4   CONCLUSIONS

A number of existing decision-support systems in the investment appraisal arena have been examined, particularly the WINA tool for whole-life costing. The main business benefits which WINA provides are greater visibility, and hence control, of costs, more accurate comparison of competing investment proposals, and improved procurement decisions. Future directions for investment appraisal, focusing mainly on the need for closer links to corporate goals and strategy, have also been addressed.

As computer power continues to increase, more powerful tools will be available. These will be supported by information from a wide range of corporate and other sources, enabling them to be used more proactively. It is believed that those organizations which succeed in the 1990s and beyond will be those that develop and exploit the information these tools provide to obtain maximum competitive advantage.

REFERENCES

1.   Paul S: 'Whole-life costing', BT Internal Report (April 1993).

2.   Gibbs J: 'Basic principles of whole-life costing', IEE Colloq on Life Cycle Costing and the Business Plan' (February 1994).

3.   Profitt T: 'Life cycle costs of aircraft systems,' IEE Colloq on Life Cycle Costing and the Business Plan (February 1994).

4.   Woodward D: 'Life-cycle costing for optimum asset management', Accounting World (February 1990).

5.   Dell'Isola A J and Kirk S J: 'Life-cycle costing for design professionals', McGraw-Hill, New York, USA (1981).

6.   Drury C: 'Management and cost accounting', Chapman & Hall (1992).

7.   Tiffin R C and Ellis H: 'Practical investment appraisal', Butterworths, London (1992).

8.   Cusack E L and Cordingley E S (Ed): 'Object oriented techniques in telecommunications', BT Telecommunications Series, Chapman & Hall (1995).

9.   Oakley P: 'Telecommunications services and the environment', BT Internal Report (October 1993).

10. Hochstrasser B and Griffiths C: 'Regaining control of IT investments', Kobler Unit Report, Imperial College, London (1990).

11. Whiting R E, Davies N J and Knul M: 'Investment appraisal for IT systems', BT Technol J, 11 , No 2, pp 193-211 (April 1993).

12. Old B: 'Corporate directors should rethink technology', Harvard Business Review, pp 6-14 (Jan-Feb 1982).

13. Roussel P A, Saad K N and Erickson T J: 'Third generation R&D', Harvard Business School, Boston, USA (1991).

14. Tate A: 'Towards the creation of a programme road map', Private Communication, Artificial Intelligence Applications Institute, University of Edinburgh (August 1992).

6

INFORMATION ACCESS FOR DECISION MAKERS

M C Revett and P R Benyon

6.1   INTRODUCTION

An essential aspect of any effective decision-making process is the availability of sufficient relevant information. Gathering and analysing information plays an important part in this process (see Chapter 1), and the assumption is that increasingly people will rely on the use of powerful personal computers (PCs) to access computer-based information sources via communications networks. Such networked systems are now seen as the source of much of the information required for a wide range of decision making within large business organizations, covering strategic, tactical and operational decisions. In addition, the view is now being taken that this approach will spread to smaller organizations and to individuals in their private lives.

The volume and diversity of information available on computer networks is growing at an accelerating rate. People are now facing the problem of information overload, which is encountered when dealing with the increasing volumes of information demanded of, and available to, them. There are significant issues in providing usable means of access to diverse sources of information held on computer networks. This chapter will discuss and review current research approaches to these issues, providing a comprehensive set of references to related work.

6.2   INFORMATION AND THE DECISION MAKER

6.2.1   The nature of information

The terms data, information, knowledge and intelligence are used in everyday speech in imprecise ways typical of human discourse. The study of the deeper meaning of these terms has been a major aspect of philosophy, particularly a branch known as epistemology, for thousands of years. In more recent times there has been renewed interest in understanding the underlying concepts, particularly in a field known as cognitive science, and its associated area of artificial intelligence (AI). The latter has been concerned with attempts to build forms of intelligence into machines, but whether this has been, or ever will be, achieved to any significant extent remains a topic of considerable controversy [1].

One view is that data is the raw material, from which information, of relevance to specific people in specific situations, can be extracted (Fig. 6.1). Aside from the information-theoretic definitions, it follows that the information content of data is critically dependent on context, and hence no absolute measure of information content can be defined. This idea underlies the economic perspective on information [2], which considers it to have potential value, either in financial or less tangible forms, that can be realized when it is acted upon. Continuing this thread, it has been proposed that knowledge is manifested only when some form of purposeful activity is carried out [3], and hence is perhaps itself only latent in information

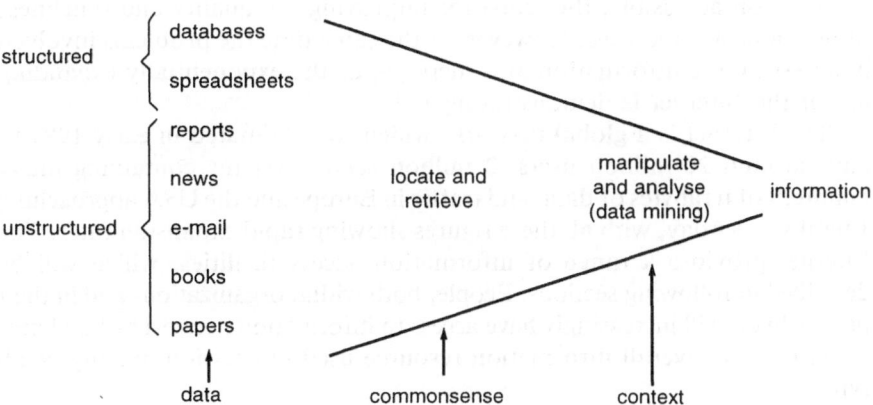

Fig. 6.1   Transformation of data into information.

extracted from computer systems. The study of these fundamental issues remains an important topic of research, and putative theories of relevance to building future information systems have begun to appear [4].

6.2.2   Information sources

The richest and most complex source of information is interaction with the natural world, via experiences, events, perceptions and emotions [5]. It is this interaction which creates much of what is often called commonsense, which underpins all human thought and decision making. Endowing machines with some form of commonsense is believed to be essential for any real form of AI, and two distinct approaches to this problem are emerging — symbolic and behavioural. An ambitious project to achieve this using symbolic AI techniques is still continuing [6], with no real indications yet of its likely success.

The second source of information is informal, unstructured information obtained via conversations, books, news reports, e-mails, etc. It also includes other forms, such as sound, images and video, and the storage and manipulation of diverse forms of information, commonly termed multimedia, present a whole range of additional problems and opportunities, which will only be considered to a limited extent in this chapter.

Computer and communications networks are increasingly becoming the main sources of computer-based information. The implicit assumption is that as the size of the network grows, increasing the amount and diversity of information accessible, the scope for improving the quality and timeliness of decisions will increase. However, at the same time the problems involved in accessing the information also increase, as the exponentially expanding use of the Internet is demonstrating [7].

The Internet is a global network, which was estimated in early 1994 to have around 20 million users, 2 million server systems containing many hundreds of terabytes of data, and traffic in Europe and the USA approaching 1 terabyte per day, with all these figures showing rapid expansion rates. The Internet provides a range of information access facilities, which will be described in following sections. People, both within organizations and in their private lives, will increasingly have access to information networks of all kinds as part of the overall information resource used in decision making of all types.

6.2.3   Use of information

In using information for decision making, a number of activities are undertaken:

- search and retrieve;
- view and manipulate;
- analyse, organize and annotate;
- edit and create.

The first two, covering information access, are the main topic of the rest of this chapter. The second two are largely synthesis activities, and have their own set of requirements for computer support tools. For use on a personal level some form of personal information manager (PIM) is required [8], although recent ideas have extended this into the concept of a dynamic personal database or 'dynabase' [9]. For use within groups, the requirements are being increasingly met by groupware applications, in research systems such as Object Lens [10] and commercial systems such as Lotus Notes [11].

As already discussed, data can be processed to yield information, and this is the principle underlying data mining (see Chapter 2).

6.3   USER INTERFACES FOR INFORMATION ACCESS

6.3.1   Background

As the means of accessing computer-based information becomes more widely available, both at work and in the home, an issue of crucial importance is the usability of the human/computer interface. While significant advances have been made in this area, particularly with the use of graphical user interfaces, it is likely that considerable developments will be required to encourage the public at large to be prepared to access computer based information. In many cases people will have limited 'computer literacy', and will make occasional and possible reluctant use of such systems, and thus the interfaces they use must be designed accordingly. In general terms, the approach being adopted is to increase the ease of use and flexibility by building a greater degree of 'intelligence' into the interface.

6.3.2   Conversational and direct manipulation interfaces

The original user interfaces to computer systems adopted a conversational or language-based model of interaction, in which the user entered a command, and the system responded. This form of interface was used on mainframe and minicomputer systems, including all UNIX systems, and on the early IBM-compatible personal computers using the DOS operating system. In all cases the commands from the user were entered in specially developed languages, designed in most cases for easy parsing and processing by the computer, with little regard for the ease of use by the human involved. The UNIX system was notorious for the terse and arcane nature of its commands, for example 'grep' and 'awk', which could provide powerful functionality for experienced users, but were intimidating for novice or occasional users.

Research into new approaches to user interfaces undertaken at the Xerox Palo Alto Research Center (PARC) in the late 1970s [12] produced the ideas that were subsequently used commercially in the Apple Macintosh, Microsoft Windows and other systems. These are the ubiquitous graphical user interfaces (GUIs), that are often characterized by their use of windows, icons, menus and pointers, and known as WIMP interfaces. The underlying approach adopted by these systems was given the name 'direct manipulation' by Shneiderman [13], based on the fact that objects of interest, such as files and programs, are represented on the computer screen in the form of icons, which are then directly manipulated by the user, by means of mouse clicks to select, and 'drag and drop'. Such systems often make use of metaphors to relate computer-based objects and operations to objects and actions in the everyday world [14]. The basic metaphor of most current systems is the desktop, upon which documents can be displayed, moved and filed. While such metaphors can have beneficial effects on usability, in many cases mixed metaphors are used, for example being able to open 'windows' on 'desktops', and in other cases attempts to avoid deviating from a metaphor can produce overly complicated interfaces [15].

A recent paper by Buxton has suggested that further major advances in user interfaces will be necessary to meet future requirements [16]. Current research in this area has strong advocates for developments in both the direct manipulation ('interface-as-tool') and conversational ('interface-as-agent') approaches, but it is likely that future advances will combine elements of both into mixed mode systems [17].

6.3.3 Visual formalisms and information visualization

As previously mentioned, metaphors are used to define the basic style of direct manipulation systems, and in an informal way assisting the understanding of the user. An alternative approach is the use of visual formalisms [18, 19], which are diagrammatic ways of representing information that have well-defined semantics for expressing and manipulating the relations within the information. As such they provide significant aids to structuring and solving problems, as has been shown for the spreadsheet, which is the most commonly encountered visual formalism. The objective of research in this area is to invent and develop a range of visual formalisms that can be used, possibly in combinations, to display and manipulate information [15].

Related work at Xerox Parc, aimed particularly at visualizing the content of large amounts of information, has drawn on the human ability to perceive spatial relationships, and to infer structure and meaning from them [20]. By using three-dimensional and animated displays, a number of novel forms of visualization have been developed, including 'perspective walls' (Fig. 6.2(a)), which provide detail views of the data of current interest and reduced scale views of the global context, and 'cone and cam trees' (Fig. 6.2(b)), which

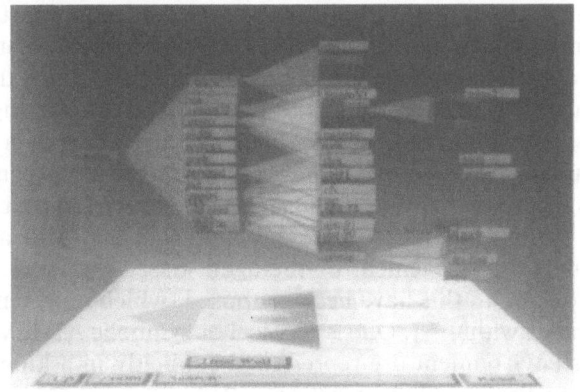

(a)

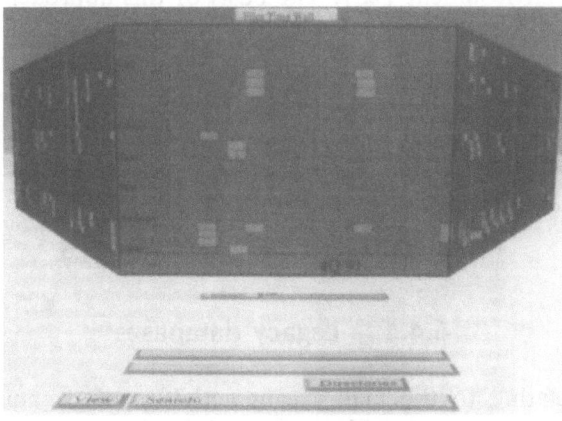

(b)

Fig. 6.2 Styles of data visualization.
[Courtesy of the Association for Computing Machinery [20], USA].

use 3D projection, shadow, and shading to display large hierarchical structures. The implementation of these ideas, because of the real-time requirements of animation, currently requires the use of specialized software techniques and expensive hardware, but technological advances will soon enable their use on standard PCs.

6.3.4   Agent-based interfaces

A development of the conversational style of user interface is the use of 'interface agents' to assist users in their task of interacting with the computer and the information being accessed [21]. Two different views of the possible role for such agents are evident in the literature, one in which the agent acts as an intermediary or mediator, interpreting and translating the user's requirements [22], and the other in which the agent acts as a third party, an assistant guiding and advising as the user interacts with the computer [23]. In both cases it is assumed that the agent has information on both the user and the task being undertaken, and the capturing and utilization of that information is a crucial element of the agent-based interface.

Two approaches to this have been proposed, which represent two major theoretical strands within AI. One is classical or symbolic AI, in which models of the agent's environment are represented explicitly in a knowledge base, which the agent uses to plan its actions [21].

A more recent development, called behaviour based AI [24], eschews the use of detailed explicit models in favour of a more reactive approach, in which the agent's defined behaviour enables it to react to its current situation and environment [25]. The basis of this approach is sometimes summarized by the phrase 'the world is its own best model'.

Work on interface agents of both types remains an active research area, but the inclusion of a form of assistant agent, given the name Intellisense, in the latest version of the Microsoft Office product, indicates that it is likely to be a topic of increasing commercial importance.

6.4   STRUCTURED DATABASES

6.4.1   Legacy databases

Most data relating to the mainstream activities of organizations, both commercial and public, is held in structured databases, which are managed by commercial database management systems (DBMS) [26]. The relational database is used for the majority of new systems developed, although many major current systems use other types of DBMS, such as network DBMS.

Databases are developed and implemented to form part of computer systems for specific operational purposes, for example, order processing, billing, accounting, work management, and for historical reasons often exist as separate systems running on different hardware and software platforms. It is widely recognized that the data residing in the databases of these legacy systems is an important corporate asset, that needs to be fully exploited if the organization is to be run in an effective and competitive manner. Many people within the organization involved in making a wide range of decisions of a short-, medium- or long-term nature could beneficially make use of data drawn from a variety of databases. It is often the integration and correlation of diverse items of data that produces the information needed for effective decision making.

6.4.2   Data warehouses

Providing convenient access to an organization's data for decision-support purposes can be problematical for a number of reasons, including:

- the need to maintain performance on vital operational systems,

- interoperation between different proprietary computer systems,

- the differing data models and standards used in the underlying databases.

An approach to solving this problem is the implementation of a corporate data warehouse [27, 28], which is a database designed and developed to hold the complete set of data relating to the operations of a corporation, and to which read-only access is provided for decision-support purposes. The data to populate the data warehouse is extracted from the operational systems at regular intervals (Fig. 6.3), enabling a running historical record to be accumulated.

6.4.3   Management information systems

The traditional means of producing management information, particularly from legacy systems, has been as paper reports, usually in standard tabular forms, produced by a regular computer run, often on a monthly basis, and delivered as a hefty computer print-out. Such reports, if they are used at all, are then often manually analysed and summarized by support people,

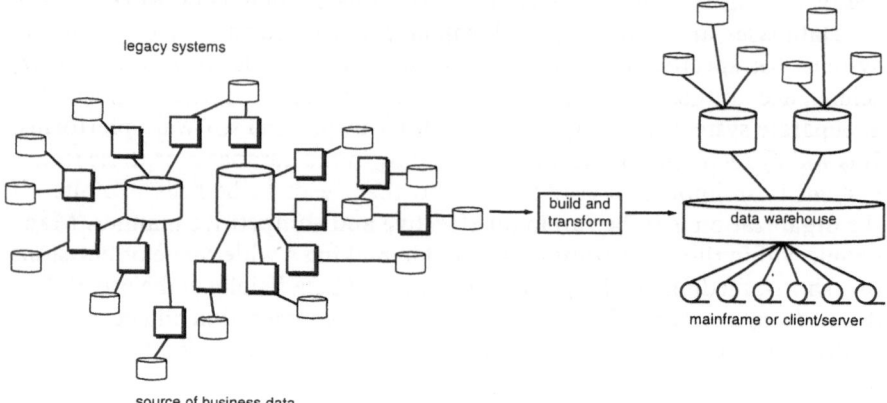

legacy systems

source of business data

Fig. 6.3   Data warehouse architecture.

before being passed to the manager who needs the information. The initial approach to providing computer-based access to such information was in effect to enable the paper reports to be displayed on screen, with some means of selecting the reports required. Management information systems (MIS) using this approach are still in widespread use, particularly in association with legacy systems, and in many cases are accessed from PCs using software that emulates a 'dumb' computer terminal.

More recent approaches to implementing MIS have been based on two computing developments, the use of WIMP interfaces, and client-server computing, in which client software, running on the user's desktop PC is supplied with data on demand by database server systems, running on remote machines and accessed via a network. For relational databases the data is obtained from the server using queries written in a standard query language (SQL [26]), which facilitates access to multiple database servers from a single client front-end.

These developments enable interactive interfaces to be built that provide users with comprehensive facilities to specify and control the content and form of the data to be displayed. One common feature is 'drill-down', in which the user 'clicks on' a number within the display and the system provides additional detail of how the number was calculated, for example drilling down on a year's sales figure for a product may display the corresponding four quarterly figures, and further drill-down can produce monthly or weekly figures. Many commercial development tools are now available for use in building GUI-based database front-ends, and flexible and usable systems can be produced.

6.4.4   Future management information system directions

The data held in MIS can be very complex, and providing means for users to search for and retrieve the data they need to address their requirements, and to understand, manipulate and display it in the form they require remains a significant topic for research. One aspect is the need to provide within the database descriptions of the data and its relationships, to enable new and infrequent users to discover for themselves the scope of the data held. One approach is to use hypertext-style interfaces (see section 6.6) to enable users to browse through descriptions of the data, and to provide integrated means of accessing and drilling down into the data [29].

Another aspect is the need for improved ways of manipulating data to create exactly the view of it required. It has been recognized that data within complex databases has a number of distinct dimensions, with associated valid ways they can be analysed, using processes such as aggregation, consolidation and summarization to create specific views. This process has been called multidimensional data analysis, and a new approach to implementing it, on-line analytical processing (OLAP), has recently been proposed [30]; commercial tools with capabilities in this area are now appearing. It is likely that the use of such techniques will place significant demands on the performance of current DBMS, and advances in the underlying technologies may be required to make their full use feasible.

6.5   TEXT-BASED INFORMATION RETRIEVAL

6.5.1   Background

The characterization and retrieval of text-based information stored on computers is an established technique [31], usually referred to simply as information retrieval, and has been widely used for many years in making bibliographic searches of on-line databases containing details of published papers. The usual approach is to base queries on Boolean combinations of keywords, which are tested against the representation of each paper stored in the database, which consists of the title, an abstract and possibly a set of index terms or keywords. Details are returned of the papers that match, in some defined sense, the combination of words given in the query. A requirement for this process to yield useful results is that the representation of each paper is accurate and comprehensive, and since titles are by nature succinct, and often can be a misleading indicator of contents, it is the abstract and index terms that are crucial. Their production is a process that requires considerable expertise and effort, usually by specialist information scientists. It has become normal practice for searches of databases to be undertaken by specialists, normally librarians or information scientists, who work in co-operation with the person wanting the information to interpret and refine the precise requirements. The process requires skill and experience on the part of the searcher, not least because of the diversity of interfaces and search techniques used by the wide range of on-line textual databases available [32].

6.5.2   Developments in text retrieval

Recent work in the area of text-based information retrieval has shown a number of trends and developments, which have been reviewed and discussed in Sparck Jones [33]. The first is the increasing availability of full text versions of documents on-line, which provide scope for improving and refining the search process. One approach is to produce a full term index for each document, which excludes all uninformative words such as 'the', 'and', 'but', 'had', etc, and to perform keyword searches against them. A more recent development of this approach is to represent each document by a vector in 'term space', which may have many thousands of dimensions, and then match queries to documents by the proximity of their term vectors in this space. The weighting of terms can be based on word counts in the documents and specified by users in the queries. The vector processing approach enables sets of related documents to be organized into clusters, and also enables a document itself to become a query, by searching for other documents that are similar to it. Commercial systems based on vector processing are available [34], and research has been reported on the use of the approach to browse through collections of documents [35], which produces a form of dynamic hypertext, as will be discussed below.

Another trend is the increasing demand for people to search, particularly on full-text documents, without the need to work through a skilled human intermediary. For this to become common practice major advances will be needed in the usability of such systems. One potentially useful technique is the use of relevance feedback [36], where the user assesses the relevance of the documents returned by initial searches, enabling the system to automatically adapt its search parameters to refine the search.

The keyword and vector processing techniques described above are based on statistical analyses of documents, with very limited concern for the structure of the text [37]. Recent research has begun to apply natural language processing and AI techniques to undertake deeper textual analysis [38], at a syntactic or semantic level, to identify concepts and topics, and a commercial product that uses these approaches has recently become available [39]. Such techniques may offer advantages in full text retrieval applications, although only limited comparisons with established methods have occurred [33]. In addition, they also offer novel possibilities for extracting information from documents, in the form of individual 'facts', which can then be combined and assembled in a variety of forms, either as summary documents or as information bases that can be used within other computer-based applications.

6.5.3   Document formats

The formats in which documents are held in computer-based systems are diverse and largely proprietary. A number of initiatives have arisen in recent years to produce standards for document formats, and the one gaining acceptance is the Standard General Markup Language (SGML) [40]. In practice SGML is a language for defining the logical structures of document types, which are known as document type definitions (DTDs). SGML is a complex and comprehensive language, and has possible applications far beyond conventional document definitions, and in many ways it represents a convergence between text-based information and structured databases, the potential of which is only now being explored.

6.5.4   Information filtering

Information filtering, in which incoming text messages, typically arriving via a computer network, are classified and routed accordingly, is related to information retrieval, although there are some significant differences [41]. Typically the process is applied to broadcast information, such as news wire services, and directed information, such as electronic messages between organizations, for example banks, or normal e-mail. One experimental application that has been reported is the filtering of news wire items [42]. The experimental use of information filtering by individuals to control and manage the information arriving via computer networks has been reported, based on the use of an interface agent which employs a genetic algorithm to adapt the filtering process to the user's current profile of interests [43].

6.5.5   Internet retrieval facilities

The Internet has a number of services that provide search facilities [7], including wide area information servers (WAIS), which were originally developed to run on a type of massively parallel computer, the Connection Machine [44]. Another is the use of 'gopher' servers, which provide a menu-driven interface to textual information held at more than 750 sites. Client machines connect to one of these servers, which effectively hide the detail of machine addresses, names, passwords, file types, etc. During a gopher session, the user navigates by selecting menu items, each of which can correspond to a file (one of six types), a directory, a search, or a remote login session on another computer. Once an interesting file has been located, the server can arrange its download to the client.

Gopher has become very popular as a distributed document handling system for Internet users. However, as more servers have been added, the problems associated with finding a particular file via menus become apparent quite quickly. To solve this problem, a database of menu item titles, called Veronica, is provided which can be searched for keywords. Gopher also provides users with index search facilities which can identify files containing certain words, although these depend upon the publisher providing such an index.

6.6   HYPERTEXT

6.6.1   Background

Hypertext has emerged as a practical way to store and organize textual information on computer systems, which facilitates the browsing and accessing of the information [45]. Where multimedia information is being accessed, the term hypermedia is normally used.'

The simplest form of hypertext provides highlighted words within the text, which when 'clicked on' produce further text to provide additional detail or explanation, or other related information. In general, a hypertext is a set of text blocks (pages or 'chunks') and an associated set of links, where each link specifies a source block and a target block, and an 'anchor' position within the source block, or both blocks (Fig. 6.4). Browsing a hypertext involves the user finding a route through the text blocks which visits those of interest.

A wide range of forms of hypertext have been developed, each with its own style and nomenclature [46]. A considerable body of research has been reported on extending and formalizing the concepts behind hypertext, and a draft standard model for hypertext, known as the Dexter model, has been produced [47, 48].

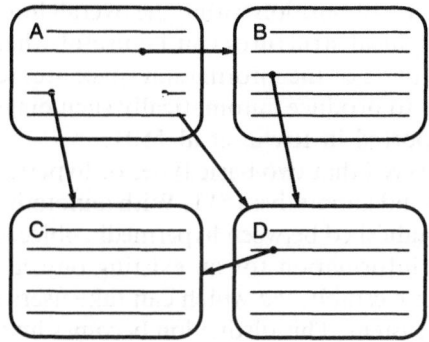

Fig. 6.4   Hypertext structure of text blocks and links.

6.6.2   Browsing

The power of hypertext lies in the ability to navigate around large information spaces, either to retrieve information on particular subjects or to browse what is there. Because the requirements of these two types of use can be quite different, hypertext systems provide several different forms of navigational aid [45]. These are important to the usability of hypertext, because a fundamental problem which can arise in navigating through hypertexts of any appreciable size has been termed 'getting lost in hyperspace' [49], and involves users encountering considerable difficulty in finding a route to, or back to, a specific page they require.

The most important browsing aids are based upon backtracking, the ability to navigate back to the previous page of information. This facility is very important for building user confidence, since it provides a simple method of 'getting back to where you were', and ultimately to return to the page where the user started. Some systems also provide history lists, which not only give users a list of nodes that have been traversed, but allow them to return directly to any one of them. A bookmark facility can also allow users to mark points to which they may wish to return.

For browsers who are new to a system, a guided tour may be provided, the object of which is to lead the user along a particular path and show them what is available. Since it is usually possible for a user to stray off the beaten track, a guided tour must have facilities which allow the user to return from any diversion. Guided tours are also provided to satisfy the needs of users with special interests who may only wish to access a subset of the information.

Overview diagrams are used to show the structure of the links and nodes at various levels of detail. The aim may be to indicate how much information is available down a particular route, to show where a user has already been, or to give users a sense of position within the overall information space. Being able to view hierarchical structures can be useful when trying to work out how detailed other parts of the information space are, relative to that already found. Techniques to produce automatically such diagrammatic navigation aids have been reported in Rivlin et al [50].

It has been observed that two basic types of hypertext links predominate, namely associative and annotative [51]. With annotative links, part-to-whole relationships are established between hypermedia objects, the links annotating or 'adding-to' the information for an existing object. In associative links, new relationships are established which can take users to entirely new areas of the hypermedia system. This distinction becomes important when building user interfaces where the use of pop-up rather than replacement windows for annotative information can help prevent user disorientation.

6.6.3   Static and dynamic hypertext

The creation of a hypertext which is easy and convenient to use for a range of users and purposes is a task requiring considerable skill and experience, particularly when this involves converting an existing conventional document [52]. At the simplest level, known as first-order hypertext, links are created that directly represent those present in the original document, for example cross-references, footnotes and bibliographic citations. This type of conversion can be automated if the document is suitably marked up, for example through the use of a standardized format defined in SGML. Second-order hypertext adds links that are not explicitly present in the original, perhaps by defining alternative paths through the hypertext for different levels of reader knowledge. A hypertext with predefined links is known as a static hypertext, and is currently the most common form.

An alternative approach, in which text blocks or links, or both, can be created automatically as the user navigates through the hypertext is known as dynamic hypertext, and related forms, with similar and additional properties, are known as intelligent hypertext or 'expertext'. Text blocks can be created automatically by extracting data from structured databases and inserting it into predefined templates, to produce a document-based report, which will always reflect the current state of the data in the database [29]. This provides similar capabilities to drill-down in MIS, as discussed in section 6.4.3. Links can be created automatically by the user highlighting sets of words within the current text block and then text-retrieval methods, as described in section 6.5.2, can be used to find the most closely related text block [35].

These approaches offer opportunities to produce hypertext that are kept in step with changing data held in other systems, and that can adapt to the requirements of the user and their task. However, even carefully designed static hypertext can present users with navigation difficulties, and thus the production of robust and usable dynamic hypertext presents considerable problems.

6.6.4 Distributed hypertext on Internet

Hypertexts have conventionally been developed by a single author or a small group, and produced in the form of a single file or directory for loading on a PC or server. In early 1993 a new form of hypertext system was introduced on the Internet, known as WorldWideWeb (WWW) [7]. It was designed to be a distributed hypertext system, in which text blocks, or pages, located on any server on the Internet could have links to other pages on any other server, irrespective of its physical location. Hence it is possible to retrieve information without having to know where on the Internet it is stored. However, unlike gopher (section 6.5.5), the user is encouraged to browse through the information via a hypertext interface, and by clicking on a highlighted area a page may be retrieved from a remote server. The standard format used for holding pages on WWW is hypertext markup language (HTML), which is defined in SGML (Fig. 6.5(a)). The system uses a standard transmission protocol, HTTP, and a standard addressing scheme, based on the use of Uniform Resource Locators, URLs. The software required to set up a WWW server is in the public domain, and hence individuals or organizations can establish a server and create their own set of pages, covering any topics they wish, with links between the pages and with other pages anywhere else on the Internet. In practice, WWW is a hypermedia system with the capability to include graphics images, sounds and video sequences within pages.

A number of GUI-based tools for browsing the WWW are also available in the public domain. The most popular of these is called Mosaic, which has versions for PCs running Microsoft Windows^{TM}, Apple^{TM} Macintoshes and systems running X-Windows^{TM} (Fig. 6.5(b)). It provides a full range of hypertext browsing facilities, including backtracking, history lists and a form of bookmark, known as a hot list. The ease of use Mosaic provides in accessing WWW has opened up the Internet to a new type of user, and it is probably the Internet service whose usage is growing fastest. Commercially supported client software, based on versions of Mosaic, is likely to become available in the near future.

```
<title>Chargecard</title>
<h1><a name="Chargecard">Chargecard<a></h1>

BT's increasingly popular Chargecard service
allows customers to make calls from almost
any phone in the UK and back to the UK from
over 120 countries around the world.

Additionally, customers can currently
make calls between
<a href="countries.html">46 countries</a>
(including within the country of call
of origin). The Chargecard can be tailored
to customer requirements, with the facility
to call anywhere or just to one number -
home or office perhaps - and have the cost
transferred to the home or office bill.
<img src="image/chargecd.gif">
<p>
```

(a)   Page description in html.

(b)   Page displayed on Mosaic.

Fig. 6.5   World Wide Web distributed hypertext.

6.7   CONCLUSIONS

A theme underlying the discussion of computer-based information access in this chapter is the considerable mismatch between the limited capacity of a typical personal computer screen to display information and the vast amounts of information that are now potentially accessible via computer networks. This chapter has considered a variety of approaches being used to address this mismatch, both in systems in current use and in research. There are two key aspects to these approaches, the first being to find effective ways of retrieving, displaying and manipulating the information at the user interface, and the second to find better ways of organizing and structuring the information on the server systems to support the effectiveness of the interface. The ideal user interface would provide uniform and seamless access to any information relevant to the task being undertaken, but current systems are far from this ideal, typically requiring users to use different interfaces, each with their own style and characteristics, to access different sources of information.

The concepts and capabilities of hypertext are likely to be of increasing relevance to computer-based information access. It is particularly suited to browsing through unstructured information, but the style of modern GUI front-ends to MIS has much in common with hypertext, and considerable potential exists for further development in this direction, including the augmenting of structured databases with descriptions of the data items they contain and the relationships between them.

The direction to be taken in developing future interfaces will be critical in addressing the problems and opportunities offered by the increasing scale of information resources available via computer networks. There are signs that a worthwhile debate is beginning to take place on the required nature and form of user interfaces, with the views of 'interface-as-tool' or 'interface-as-agent' representing two poles in the debate. In practice, interfaces usually include elements of both, and it is likely that real advances will be made when effective ways of combining the strengths of both are found.

REFERENCES

1.   Graubard S R: 'The artificial intelligence debate — false starts, real foundations', The MIT Press (1988).

2.   Saxby S: 'The age of information', Macmillan Press Ltd, pp 10-19 (1990).

3.   Newell A: 'The knowledge level', Artificial Intelligence, $\underline{18}$, pp 87-127 (1982).

4.   Devlin K: 'Logic and information', Cambridge University Press (1991).

5.   Turner M: 'Expert systems and decision support', Expert Systems for Information Management, 1 , No 1, pp 3-21 (Spring 1988).

6.   Guha R V and Lenat D B: 'Cyc — a midterm report', AI Magazine, 11 No. 3, pp 32-59 (Fall 1990).

7.   Dern D P: 'The Internet guide for new users', McGraw-Hill (1994).

8.   Kaplan S J, Kapor M D, Belove E J, Landsman R A and Drake T R: 'Agenda: a personal information manager', Comm ACM, 33 , No 7, pp 105-116 (July 1990).

9.   Press L: 'Emerging dynabase tools', Comm ACM, 37 , No 3, pp 11-16 (March 1994).

10. Lai K, Malone T and Yu K: 'Object Lens: a 'spreadsheet' for cooperative work', ACM Trans Inf Sys, 6 , No 4, pp 332-353 (October 1988).

11. Golfin N and Jackson M: 'Groupware trial in BT', BT Technol J, 12 , No 3, pp 51-55 (July 1994).

12. Smith D C, Irby C, Kimball R, Verplank B and Harslem E: 'Designing the STAR user interface', Byte, 7 , No 4 (1982).

13. Shneiderman B: 'Direct manipulation: a step beyond programming languages', IEEE Computer, 16 , pp 57-69 (1983).

14. Marcus A: 'Human communication issues in advanced UIs', Comm ACM, 36 , No 4, pp 101-109 (April 1993).

15. Nardi B A and Zarmer C L: 'Beyond models and metaphors: visual formalisms in user interface design', J Visual Languages and Computing, 4 , pp 5-33 (1993).

16. Buxton B: 'HCI and the inadequacies of direct manipulation systems', SIGCHI Bulletin, 25 , No 1, pp 21-22 (January 1993).

17. Frohlich D M: 'The history and future of direct manipulation', Behaviour and Information Technology, 12 , No 6, pp 315-329 (1993).

18. Harel D: 'On visual formalisms', Comm ACM, 31 , No. 5, pp 514-530 (May 1988).

19. Shneiderman B: 'Beyond intelligent machines: just do it', IEEE Software, pp 100-103 (January 1993).

20. Robertson G G, Card S K and Mackinlay J D: 'Information visualization using 3D interactive animation', Comm ACM, 36 , No 4, pp 57-71 (April 1993).

21. Chin D N: 'Intelligent interfaces as agents', in Sullivan J W and Tyler S W (Eds): 'Intelligent User Interfaces', ACM Press (1991).

22. Weiderhold G : 'Mediators in the architecture of future information systems', IEEE Computer, pp 38-49 (March 1992).

23. Maes P: 'Learning interface agents', Proc ACM SIGCHI Int Workshop on Intelligent User Interfaces, Orlando, Florida (1992).

24. Maes P: 'Behavior-based artificial intelligence', Second Conf on Adaptive Behaviour (December 1992).

25. Maes P: 'Agents that reduce workoad and information overload', Comm ACM, 37 , No 7, pp 31-40 (July 1994).

26. Date C J: 'Introduction to databases systems, Vol I', Addison-Wesley (1990).

27. Inmon W H: 'Building the data warehouse', QED Information Sciences (1992).

28. Eberhard R, Kidd W, Martin J, and Moseley M: 'Information warehouse: a user experience', Info DB, 7 , No 2, pp 6-12 (Spring 1993).

29. Beiber M: 'Automating hypermedia for decision support', Hypermedia, 4 , No 2, pp 83-110 (1992).

30. Codd E F, Codd S B and Salley C T: 'Providing OLAP (on-line analytical processing) to user-analysts — an IT mandate', E F Codd & Associates (1993).

31. Salton G: 'Automatic text processing — the transformation, analysis and retrieval of information by computer', Addison-Wesley (1989).

32. Vickery B and Vickery A: 'Online search interface design', J of Documentation, 49 , No 2, pp 103-187 (June 1993).

33. Sparck Jones K: 'Assumptions and issues in text-based retrieval', in Jacobs P S (Ed): 'Text-based intelligent systems: current research and practice information extraction and retrieval', Erlbaum Lawrence Association, pp 157-177 (1992).

34. McGonigle D and Golly L: 'Lost in space? — an intelligent retrieval solution', AI Expert, pp 50-53 (June 1993).

35. Salton G, Allan J and Buckley C: 'Automatic structuring and retrieval of large text files', Comm ACM, 37 , No 2, pp 97-108 (February 1994).

36. Salton G and Buckley C: 'Improving retrieval performance by relevance feedback', J Am Soc Inf Sci, 41 , No 4, pp 288-297 (1990).

37. Stanfill C and Waltz D: 'Statistical methods, artificial intelligence and information retrieval', in Jacobs P S (Ed): 'Text-based intelligent systems: current research and practice in information extraction and retrieval', Erlbaum Lawrence Association, pp 215-225 (1992).

38. Sparck Jones K: 'The role of artificial intelligence in information retrieval', J Am Soc Inf Sci, 42 , pp 558-565 (1991).

39. Oracle Corporation: 'ConText', Oracle White Paper (September 1993).

40. Bryan M: 'SGML — an author's guide to the Standard Generalized Markup Language', Addison-Wesley (1992).

41. Belkin N J and Croft W B: 'Information filtering and information retrieval: two sides of the same coin', Comm ACM, 35 , No 12, pp 29-38 (December 1992).

42. Lehnert W and Sundheim B: 'A performance evaluation of text-analysis technologies', AI Magazine, 12 , No 3, pp 81-94 (Fall 1991).

43. Sheth B and Maes P: 'Evolving agents for personalised information filtering', Proc Ninth Conf on Artificial Intelligence for Applications (CAIA-93), IEEE Press, pp 345-352 (1993).

44. Stanfill C and Kahle B: 'Parallel free-text search on the connection machine', Comm ACM, 29 , No 12, pp 1229-1239 (December 1986).

45. Nielsen J: 'Hypertext and hypermedia', Academic Press (1990).

46. Rada R: 'Hypertext — from text to expertext', McGraw-Hill (1991).

47. Halasz F and Schwartz M: 'The Dexter hypertext reference model', Comm ACM, 37 , No 2, pp 30-39 (February 1994).

48. Gronbaek K and Trigg R H: 'Design issues for a Dexter-based hypermedia system', Comm ACM, 37 , No 2, pp 40-49 (February 1994).

49. Edwards D M and Hardman L: 'Lost in hyperspace: cognitive mapping and navigation in a hypertext environment', in McAleese R (Ed): 'Hypertext theory into practice', Intellect (January 1989).

50. Rivlin E, Botafogo R and Shneiderman B: 'Navigating in hyperspace: designing a structure-based toolbox', Comm ACM, 37 , No 2, pp 87-96 (February 1994).

51. Bielawski L and Lewand R: 'Intelligent systems design', Wiley (1991).

52. Rada R: 'Converting a textbook to hypertext', ACM Trans on Inf Sys, 10 , No 3, pp 294-315 (July 1992).

7

A KNOWLEDGE-BASED SYSTEM FOR THE CONFIGURATION AND PRICING OF NETWORK MANAGEMENT SYSTEMS

D L Scott and P M Bull

7.1 INTRODUCTION

Engineering design is a complex process requiring solutions that meet the customer's requirements within the constraints of design rules and cost. The initial phase of the design process usually requires a combination of creativity and expertise to translate the requirements into a technically feasible solution. While it would be extremely difficult to build creativity into a tool, providing active support for the designer can help to ensure that constraints are not overlooked. Later, the detailed design generates a range of outputs such as parts lists and cost breakdown. These tasks are laborious and potentially error-prone and thus suited to automation.

This chapter describes a tool developed for engineers designing network and service management systems [1] in communications networks using BT's ServiceView™ product. The ServiceView (formerly Concert) expert configuration and pricing tool (CECPT) features a graphical interface for the entry to the network design schematic, supported by expert systems technology to address both of the areas identified above. An overview of the network management system and the design process are given in section 7.2, followed by a discussion of some aspects of the design problem supported by CECPT. Development of the tool is discussed in section 7.4, including practical considerations involved in developing a supportable knowledge-based application. An unusual aspect of the development is the use of commercial Windows™ products as functional modules integrated to operate as a single application. The benefits and disadvantages of this approach are discussed.

7.2 BACKGROUND

The management of customer communications systems may be undertaken by the network operator, the customer or a third party. The types of management features required define the structure of the management system to be used. There are three levels of system management (see Fig. 7.1) that may be deployed:

- network management system (NMS) — concerns the management of the core network;

- service management system (SMS) — monitors and controls the services available on the network;

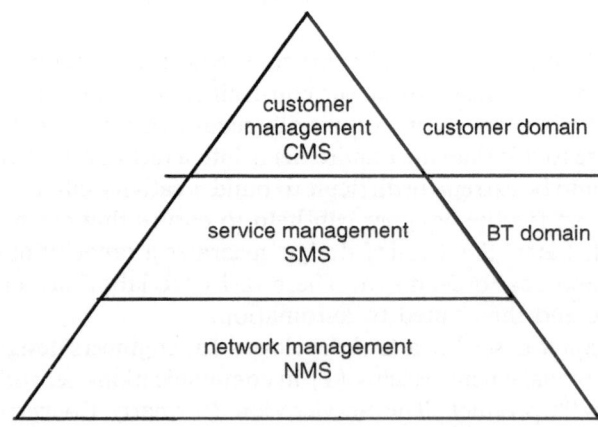

Fig. 7.1   ServiceView deployment levels.

- customer management system (CMS) — gives the customer a view of the supplied managed services and allows the management of internal private networks. At this level the customer may use the system for monitoring network performance, fault reporting, service changes, ordering and billing, etc.

The management system may be for a large national or multinational network spanning many sites and countries. A schematic diagram for a relatively simple three-layer management system is shown in Fig. 7.2. It consists of server systems which store information about the network configuration and receive, log and process network events such as faults or changes. The servers are designated CMS, SMS or NMS according to the layer at which they are operating. Access to this information is through one

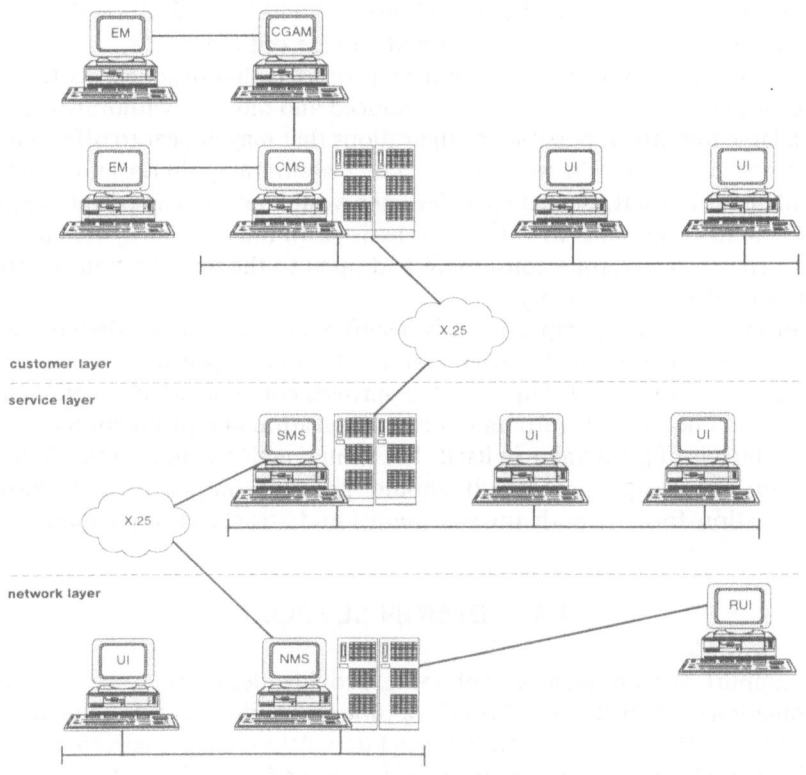

Fig. 7.2   Schematic of a management system.

or more user interface (UI) systems at the network control centre(s). A UI that is not co-located with the server system but is connected to it via a low-speed link is termed a remote user interface (RUI). The management system may interact with communications equipment via its proprietary management system, known as an element manager (EM).

The schematic is a top-level view of the design; each of the component systems must now be dimensioned and their hardware and software configurations determined. Configuration of a server, which is one of the more complex components, will be used to illustrate the problem. The hardware platform type (e.g. Sun Sparc 2^{TM}, Sparc 10^{TM}) will be determined by a number of factors including the size of the network, the number of user interfaces connected to it, anticipated alarm rates and the management services to be supported. The configuration of this platform in terms of memory, disk space, etc, must then be determined. Additional hardware may be required, for example to connect to an X.21 link. The server's software consists of a range of modules, including some which are platform dependent and some required to support selected service options.

A complication in the configuration process is that products evolve over time, with new components being introduced and old ones withdrawn. There are a large number of possible configurations that may appear to offer similar solutions, but some do not meet product assurance criteria and must be avoided. As a result, the network designer would have to call on the support of a technical specialist to validate or assist with the detailed system design. This process can be time consuming and open to the introduction of errors if it is performed manually.

From a business perspective it is beneficial to provide the designer with a tool that embodies the knowledge of the technical specialist and financial authority and which holds up-to-date approved component information. The tool should not merely automate the configuration and production of parts lists, which is of great value in itself, but should actively support the designer through the design process. It should produce output in a consistent presentation format, both for schematics and structured reporting.

7.3 DESIGN SUPPORT

To support the designer a tool must provide design-rule checking and automation of tasks that are difficult or time consuming to perform manually. By reducing the time spent on low-level detail the designer can concentrate on high-level design issues and investigate alternative solutions. Some aspects of the design problem are discussed below.

7.3.1 Management system performance requirements

The number of items in the managed network that are capable of issuing alarms will have a direct impact on the size of the management information base (MIB) required on the server. The event rate, both static and burst, determines the processing power and memory requirements of the server. The server dimensions may be modified, dependent on the number of users of the system, the types and frequency of queries, or network views requested.

The low-speed communications link associated with an RUI restricts information transfer with the server. To maintain performance for the operator the RUI must have increased storage and processing capacity.

The number of operators accessing the server also has an impact on the performance of the system, including processing power of the server and traffic on the LAN. The system designer has to consider these factors when determining the server dimensions and LAN topology.

7.3.2 Upgrade considerations

A management system is likely to undergo significant changes during its life as the customer's communications infrastructure evolves and new technology becomes available. A configuration tool must therefore be capable of determining the incremental differences between an existing installed system and a variety of possible migration options. For example, a customer may decide that resilience of the communications infrastructure should have a higher priority than that for the existing installation. This places a higher demand on the availability of the management system. To increase reliability it may be necessary to duplicate components such as the server, power sources, communications links and storage devices.

7.3.3 Software licensing issues

ServiceView uses a variety of third-party software products. The licences for these products may be based on the platform performance or the number of users. The designer must ensure that the correct licences are specified for each configuration.

7.4 CECPT SYSTEM DEVELOPMENT

The development approach and environment were influenced by a number of factors. A working prototype was required to demonstrate feasibility.

Secondly, although the CECPT was expected to be installed at design centres in the UK and at designated centres throughout the world, the total number of systems to be deployed was not expected to be large. This meant that usability had a high priority and user training requirements were to be minimized. International use meant that the tool would have to cope with country-specific variations of configuration.

The user interface is a key area of the development because it not only determines the usability of the tool, but it also consumes a significant proportion of the development effort. Two methods of data entry were considered.

• Forms entry — this would require a set of standard templates that the designer would fill in to describe the network requirements. Each workstation would need to be specified together with the connectivity between the devices. There are a number of disadvantages with this approach. Data entry becomes a rather laborious process. It also becomes difficult to view the network in its entirety to ensure that all the connectivity has been entered. The system designer would probably have to refer to a drawing or sketch of the network.

• Schematic entry — this uses a drawing showing symbols and associated interconnectivity as the design master. This technique is widely used in computer-aided design, for example to enter graphical descriptions of integrated circuit and printed wiring board designs. This is a more natural way for a designer to work.

Since a schematic of the network is needed at various stages of the design, the latter approach was adopted.

7.4.1   Architecture

The component parts of the application are shown in a block diagram in Fig. 7.3. The systems engineer has access to:

• the security module, providing password access to diagrams and reports — this has the facility to allow different access levels dependent on the role of the user (i.e. bid manager, system designer, financial authority), and these levels may be used to change the view of, or filter, the available data;

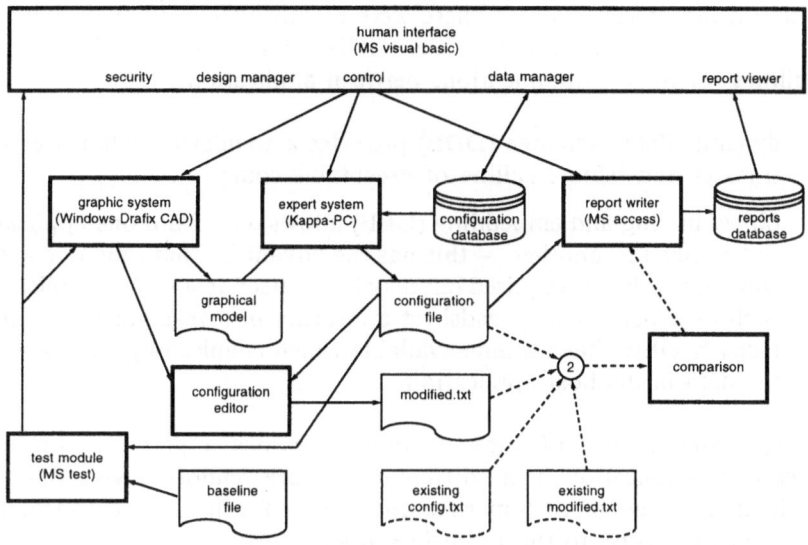

Fig. 7.3   Architecture block diagram.

- the design manager, enabling the user to set up a hierarchy of folders, projects and jobs — this removes the need to understand the directory and file structures of the host computer and provides 'housekeeping' facilities for designs such as copy, rename, move, delete, etc;

- a data manager, giving access to the information held in an external database, as shown in Fig. 7.3, allowing a system administrator to add or modify configuration details and pricing information;

- the report viewer, providing a screen display of the output reports;

- the drawing area, allowing the designer to enter, annotate and modify a schematic of the network management system.

The design process begins with the entry of the initial diagram. The graphical model is transferred to the expert system which creates an internal representation. The expert system then builds a hierarchy of physical system objects, such as servers and UIs, and configures each with information extracted from the external database or from the rules held internally. A configuration file is transferred to the report-writing system for formatting and storage. When the process is complete, the user may view or request hard copy of the reports using the report viewer.

Consideration had to be given to the office automation equipment already in use in the design centres. These were mainly PCs. Microsoft Windows provides a graphical user interface (GUI) for the PC and also provides facilities for intercommunications between applications:

- dynamic data exchange (DDE) provides a standard for bothway data transfer and remote calling of executable code;

- object linking and embedding (OLE) allows data from one application to be linked to another — this has the advantage that only one master copy of the data is required but may be accessed from other applications such as a part of a spreadsheet appearing in a text document; more recently OLE2 has become available which enables in-place editing of the data in the host application.

The combination of these facilities provides a good platform for integrating applications from different software vendors, giving flexibility to choose packages on their merit as appropriate for the task. The main tasks were partitioned on to the following sub-systems:

- graphical entry system;
- expert system and object model;
- report generation facilities;
- design manager;
- data manager.

7.4.2   Graphics sub-system

Schematic entry of the management network requires a drawing package with facilities for interrogating the design and extracting the topology and symbol attributes. Most standard drawing packages do not have these advanced facilities which are normally found in more expensive CAD (computer aided design) tools. A disadvantage of fully featured commercial CAD packages is that they are usually difficult to learn and may be targeted at particular functions such as mechanical design or electronic circuit layout. Two key criteria for selection of the CAD package were:

- support for object-based drawings;
- a good built-in programming or macro language.

The chosen package, Drafix Windows CAD^{TM}, is an object-based drawing tool that allows the use of custom symbol libraries and provides general line and polygon drawing functions. Each item in the drawing is handled as an object with associated attributes, for instance a line has attributes for colour, width, angle, start and end points. Each object also has a unique identifier (UID) to distinguish it from all other objects in the drawing. User-definable attributes may be added to any of these objects (i.e. cost, owner, etc). The tool also has an extensive built-in macro language which can be used to interrogate each object and extract or modify the attribute values.

In order to reason about the schematic, the model has to have knowledge of the connectivity between the system components. This implies that the drawing package must create and maintain the connectivity information. This facility was not directly supported by Drafix but was implemented using its macro programming language. Attributes added to each of the connecting lines hold the values of the symbol UIDs connected to each end. Each symbol has an attribute that stores the UIDs of all lines connected to it. These attributes are updated whenever new connections are made between symbols.

7.4.3   Network editor

Replacing the tool bar and menu options of the standard graphics package with those appropriate for designing ServiceView systems provides the network editor as shown in Fig. 7.4. Starting a new design creates a drawing template and prompts for details such as design name, customer, etc, which is then displayed on the lower right of the drawing. A tool bar of picture buttons along the top of the window gives access to drawing functions such as placing and connecting ServiceView symbols.

7.4.4   Correct by construction

The graphics package places no restriction on the placement of symbols or their connectivity. To provide a mechanism to check the correctness of the design, a set of rules was developed to exclude certain combinations of components (say attaching a CD-ROM drive to a printer). These rules are implemented as a set of all valid component types that may be connected, which implies that if a symbol pair is not in the list then the connection is invalid.

Checking for valid symbol connections during the drawing stage has the advantage that the designer gets immediate feedback on a problem rather than waiting for warning messages during the configuration process. It also

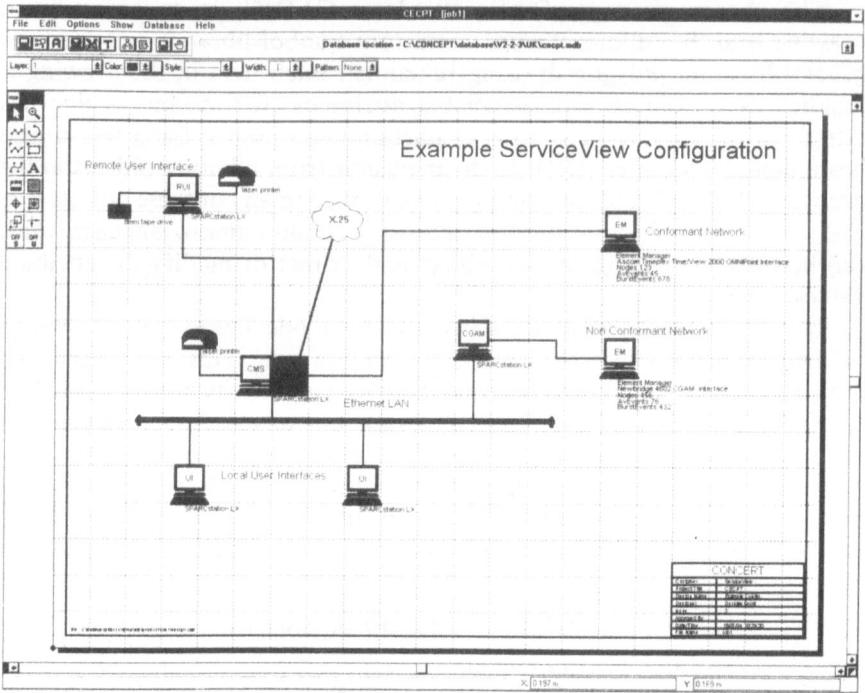

Fig. 7.4   Network editor.

reduces the configuration complexity as only valid connections are processed, requiring less error checking during the configuration phase. This has the additional benefit that adding new symbols to the tool, as its functionality is increased, can be accomplished in a controlled manner.

7.4.5   Symbol attributes

Information may be required to specify additional applications to be used on a server such as fault-monitoring or terminal-emulation packages. In the case of an element manager the type of network being managed must be specified in order that the appropriate software is added to the configuration. To enable the performance requirements of the server and LAN to be calculated, the size of the network, and expected event rate from it, must be specified. CECPT provides a context-sensitive menu for editing these attributes. The menu for a network element manager is shown in Fig. 7.5.

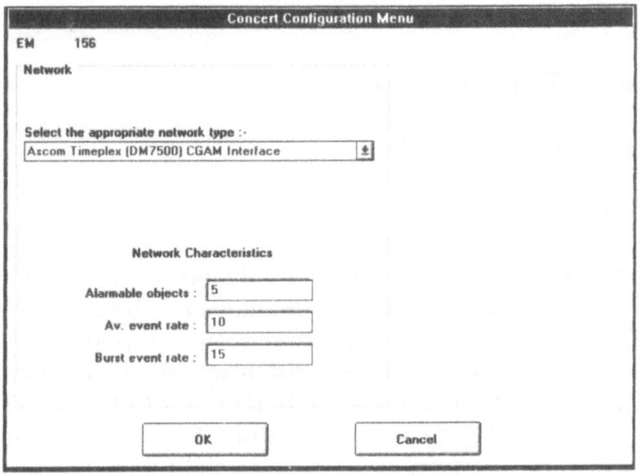

Fig. 7.5   Element manager attribute menu.

7.4.6   Expert system

Why use an expert system (ES)? Configuration tools are frequently developed using conventional software and database techniques. They tend to be quantitative in nature: 'If part number is X002, then add part number Y040'. In contrast, designers make extensive use of qualitative (or 'heuristic') factors: 'If the number of elements managed is large and the alarm rate is high, then upgrade the server to a more powerful processor or add more memory'.

One ES approch to a configuration type problem is to use a rule-based system. A number of rules are defined in the form of 'IF ... THEN ... ELSE' statements, for example:

Rule 1:

IF
   Processor = 'Sparc IPX'
AND
   Performance = Low
THEN
   Processor = 'Sparc 2'

Rule2:

IF
    NumberOfElements = Large
AND
    AlarmRate = High
THEN
    Performance = Low

However, rule-based systems have disadvantages for anything beyond small systems. As a rule base grows, maintenance becomes a significant problem. The development team spend a greater proportion of their time checking for rule conflicts and inconsistencies. A commonly cited (and successful) example of such a system is DEC's XCON [2, 3] that was developed to configure VAX computer systems. As this system grew to about 5000 rules it became necessary to restructure and rationalize the rule base. Later, as maintenance effort grew again, a system was built to assist with rule-base maintenance and to perform validation. In DEC's case, there was a reported 40% product churn each year, with a corresponding level of rule-base change. It required full-time effort of several knowledge engineers to maintain the system. An important requirement for the CECPT was that its knowledge should be maintainable by authorized users, probably non-programmers, as far as possible.

An alternative approach for representing knowledge, proposed by Minsky [4] and known as 'termed frames', overcomes some of the limitations associated with rule-based systems. This technique has been adopted by AT&T in a system for product configuration [5] in which the need for the knowledge base to be maintained by non-programmers is also highlighted.

Configuration is only part of the CECPT's task. An additional objective was to provide a design support tool sympathetic to the way designers work. Such support may be in the form of offering the designer only those options that are allowable in the current context or to warn that items have been inappropriately connected together. The natural approach of a designer to a problem is to start by drawing diagrams of the system in which real-world objects are represented in relation to each other, in effect creating a model of the system. The designer then reasons about the model in terms of the behaviour of the objects. Model-based reasoning (MBR) is founded upon this approach and there are a number of commercial development tools available. The model consists of objects (e.g. servers) that have attributes, such as 'server-type' and 'connected-to' (which is a list of objects connected to it). Behaviour is modelled by means of messages and methods. For example, a server could ask its element managers to tell it their alarm rates, etc, from which the server can deduce its own processor and memory requirements.

A hybrid system combines model- and rule-based approaches to draw upon the strengths of both. Object oriented programming is used to represent the behaviour and attributes of individual objects. Rules are used to provide IF-THEN-ELSE reasoning that may, or may not, relate to specific objects. Because a number of small rule-bases are held to address specific sub-problems, the tool becomes much more maintainable.

The expert system shell used, Kappa-PC^{TM}, provides an object oriented system with a rule-based facility. These are used together to produce a hybrid system. The initial stage of configuration consists of transferring the schematic drawing information into the expert system. An object model of the symbols and relevant attributes is created in memory. This is used to trace any connections between the objects, such as the number of UIs attached to a server, or find the host of a remote UI. From this a 'physical' system model is constructed, which enables unique instances of system types to be created and configured (i.e. hardware and software components added). An extract of the class and instance hierarchy is shown in Fig. 7.6.

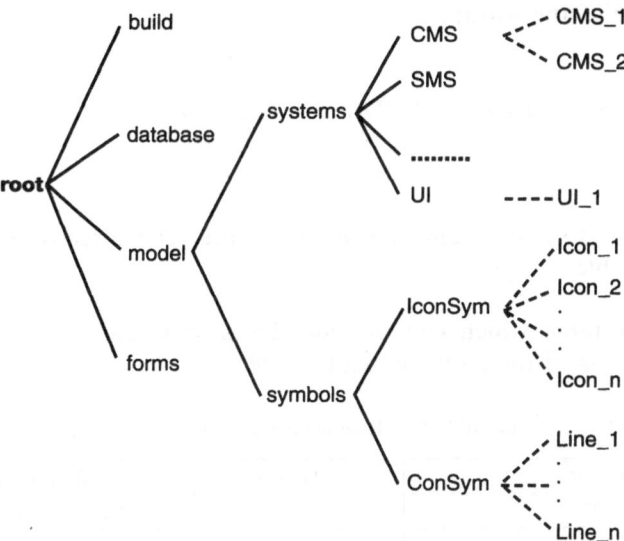

Fig. 7.6   The object hierarchy.

An example of a configuration rule to upgrade a workstation from a 'SPARCstation IPX' to a 'SPARCstation 2', if the number of users is greater than 1, is shown below:

    IF
        Processor = 'SPARCstation IPX' AND Users > 1
    THEN
        Processor = 'SPARCstation 2'

Similarly, a SPARCStation 2 might be upgraded to a SPARCStation 10, if the number of users is greater than 10:

    IF
        Processor = 'SPARCstation 2' AND users > 10
    THEN
        Processor = 'SPARCstation 10'

It should be noted that the structure of the two rules is similar — only the specific details differ. This can be exploited to generalize the more commonly used rules and 'table-drive' them at run time from a database. These rules are of the form:

    IF
        (Processor = *processor* AND users > *users*)
    THEN
        (Processor = *newprocessor*)

where *processor, users* and *newprocessor* are read directly from the database table.

A database table which can provide the specific parameters to the generalized rule is of the form shown in Table 7.1.

Table 7.1   Rule database table.

Processor	Users	Newprocessor
SPARCstation IPX	1	SPARCstation 2
SPARCstation 2	10	SPARCstation 10

A module called 'Data' maintains the database tables. By this means an authorised user, rather than the developers, is able to maintain the most frequently changing system knowledge. The use of generic rules also considerably reduces the size of the rulebase. Where knowledge is far more stable and specific, it is more efficient to code the rules directly into the rulebase.

The rules for checking connectivity are derived directly from a drawing containing all the valid symbol connections. This has the advantage that it provides a visual record of all legal connections and is more easily updated than a textual record.

7.4.7   Testing

It is important to be able to test a multi-application tool, such as described, but a variety of problems were encountered during the development:

- each of the constituent application sub-systems has its own non-standard proprietary language;
- traditional support tools for optimizing, testing and measuring code metrics are not readily available;
- the system is event driven, making the permutations of input characteristics difficult to obtain.

The approach taken in testing the tool has been to produce report sets from several target designs which have then been reviewed by experts and to use these as fixed baseline reports. The baseline data can easily be compared to the new reports and any differences found are highlighted. A simple file comparison program is all that is required to compare the output reports. Both the configuration details and the financial summary information are checked after each test run, providing a means of regression testing of the system during development and before delivery to ensure consistency with previous releases. The drawback with this approach is that the functionality of the graphics sub-system or the symbol libraries is not exercised.

To test the user interface and graphics system a recording technique is used where user key-strokes and mouse clicks are stored. The resulting file may be repeatedly played back and resulting configuration files compared with the baseline.

7.5   BENEFITS

There are a number of benefits provided by CECPT, including the following.

- Consistency of presentation of diagrams and output reports — historically designers have used a multitude of drawing packages each with their own set of symbol libraries and layout formats. This leads to inconsistencies and is open to misinterpretation.

- Use of a managed database of third-party hardware and software components with up-to-date pricing information and a consistent numbering system.

- Correct by construction methods — the tool enables the user to produce only 'legal' schematics through context-sensitive menus and critiquing against design rules. The automated build process ensures that the output configurations and costings are consistent with the schematic. Where the tool determines that, say, a processor needs to be upgraded, the schematic is automatically back-annotated accordingly.

- The tool supports a design and business process, which is important from a quality perspective. Moreover, a consequence of developing a tool such as this is that the process itself can be refined.

- Embedding expertise in the system reduces the learning curve for new designers. It also reduces vulnerability to the loss of key skills.

The system enables a rapid response to an 'invitation to tender' and allows a more flexible approach in that a variety of design topologies may be considered and costed quickly. Productivity of system designers is increased and allows more effort to be focused on the actual design problem. The customer perception is of a responsive, flexible supplier.

As expertise is gained in the design task, the rule set may be extended to cover heuristic information or even real examples of past designs.

7.6   CONCLUSIONS

This chapter has described the development of a tool used for assisting a network designer during the bid phase for a management system. It has been shown that it is possible to integrate applications within the Microsoft Windows environment. The application architecture for the tool was chosen for its flexibility, with the use of a commercial CAD package for data entry and a rule-/object-based expert system for the creation of the system object model. An advantage of using this particular set of applications is that they provide an interpreted development environment that enables rapid modification and testing of code without the continual need for compilation, although this does have an impact on overall system performance.

Although a particular tool has been described, this is only a single implementation using the above applications architecture. The same approach can be applied to many other configuration tasks where a complex process can be broken down into a combination of algorithms and heuristic expertise. The architecture described has also been applied to the design of building wiring systems. Examples of other potential applications include configuration of LANs, PBXs, private networks, etc.

At the time of writing, the CECPT has been deployed at a number of sites within the UK and is expected to be deployed to other sites around the world.

REFERENCES

1. Special issue on: 'Network management systems', BT Eng J (October 1991).

2. McDermott J: 'R1: a rule-based configurer of computer systems', Artificial Intelligence, 19 , No 1 (1982).

3. Barker V and O'Connor D: 'Expert systems for configuration at digital: XCON and beyond', Communications of the ACM, 32 , No 3 (1989).

4. Minsky M: 'A framework for representing knowledge', Psychology of computer vision (1975).

5. Wright J R et al: 'A knowledge-based configurator that supports sales, engineering and manufacturing at AT&T network systems', AI Magazine (1993).

8

A NEW APPROACH TO CONTRACT SPECIFICATION

M T Last and S L Corley

8.1   INTRODUCTION

The task of producing and reviewing contracts — or service level agreements (SLAs) — requires a detailed knowledge of the business processes, business drivers (e.g. costs, efficiency) and co-operating business units. In a large organization it is impossible for a small number of individuals to carry out this task, so responsibility for SLA specification is generally distributed. Differences in approach to the task can result in inconsistencies in the way measures are used within similar SLAs. One consequence of this is that a large number of unnecessary performance measures may be created. Mechanisms for monitoring, storing, processing and maintaining each of these additional measures have to be implemented and maintained, and large amounts of extra data therefore have to be handled. Additionally, changes to the process, the organizational units and particularly the business drivers can impact on the measures chosen for a specific SLA. There is, therefore, a need to:

- minimize the time and effort required to produce SLAs;

- create and use measures consistently;

- minimize the measures required for each SLA;

- maximize the flexibility of SLAs with respect to the business drivers.

This chapter describes a technique called case-based reasoning, which may help in addressing these challenges. The technique has been used to implement a demonstrator decision-support system to assist SLA specifiers in selecting measures.

8.2   SELECTING MEASURES FOR SLAs

Within a customer-supplier relationship, there are many aspects of the supplier's performance which could be measured using an SLA. However, the creation of unnecessary measurement data wastes time and money. Each SLA should, therefore, contain the minimum set of measures which allow the supplier's performance to be assessed. One reason for disregarding a particular measure may be that it has minimal impact on the service provided to the customer. Another may be that it is not related to the business drivers impacting the customer. Business drivers are the internal or external influences on a company's financial performance [1]. There are many drivers, including regulation, competitive pressure, quality, customer satisfaction, and of course costs. The relative importance of these business drivers has a major influence on how the customer unit is assessed, and hence on which aspects of supplier performance should be measured.

8.3   CASE-BASED REASONING

Case-based reasoning (CBR) is a technique which provides the ability to use solutions generated for previous problems to help guide in the solution of a new problem [2]. Each previous experience (or case) can contain a great deal of information including the characteristics of the situation encountered (e.g. a description of the situation, how it differs from similar situations, etc), the approach taken to solving it, the results, and whether the outcome was good or bad. Each of these items of information is known as a feature. CBR depends on being able to characterize every problem as a set of such features. The features of a new problem are matched against each problem in the case base (the store of previous experiences), in order to determine which of the previous solutions are likely to be the most relevant.

The basic processing cycle for CBR (Fig. 8.1) is as follows:

- input a description of the problem;

- extract the important characteristics (features) of the problem;

- retrieve relevant past solutions based on the problem characteristics;

- adapt the past solutions to the current problem;

- evaluate the proposed solutions;

- execute the best solution and analyse the results;

- store the new case together with its solution.

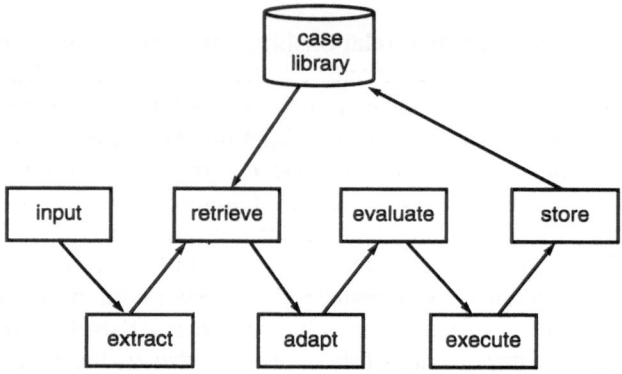

Fig. 8.1   CBR processing cycle.

CBR has been used to address many problem-solving tasks including planning [3], diagnosis [4] and design [5]. The approach has a number of benefits, including:

- solutions are possible in new or poorly-understood domains;

- system performance improves with experience;

- 'bad' solutions can be used to avoid repeating previous mistakes;

- solutions can be produced even with incomplete or incorrect information.

8.4   APPLYING CBR TO MEASURE SELECTION

CBR was chosen for the measure selection system because:

- the rules for selecting appropriate measures for SLAs are poorly defined;

- a large number of examples of the results of applying expertise to the problem already exists (i.e. existing SLAs);

- it is desirable for the system to be able to improve with experience and to benefit from the experiences of other SLA developers (as a human expert does).

In applying CBR to measure selection, it is assumed that both the business drivers and the characteristics of the customer/supplier relationship influence the choice of measures. In general, similar measures should be chosen under similar business conditions (e.g. the same types of measure should be selected for any SLA for which the key driver is cost reduction). Also, similar measures should be chosen for similar customer/supplier relationships (e.g. the same types of measure should be selected for all repair services). The degree of emphasis to be placed on each of these influences, and others, can currently only be determined by experts. However, existing SLAs contain measures which have been chosen by experts to reflect both of these influences. CBR, therefore, provides the capability for the expertise which is implicit in these existing SLAs to be used in guiding the selection of measures — either for a new SLA or for an existing SLA when business conditions change.

Proprietary tools are available which provide CBR engines for comparison of cases. These allow the most closely matching cases to be identified. However, much of the intelligence of the measure selection system resides within the software which has been developed to allow adaptation of existing measures. The various components of the system are described in the next section.

8.5   IMPLEMENTATION

8.5.1   Case structure

Each SLA in the system is represented by a case in the case base. The structure of the cases is illustrated in Fig. 8.2.

feature	value
service description	text
customer	unit name
supplier	unit name
measures	list of measures
importance of QoS	[1-10]
importance of cost	[1-10]
importance of quality of work	[1-10]
importance of efficiency	[1-10]
importance of revenue	[1-10]
importance of market share	[1-10]

Fig. 8.2   SLA case structure.

The first feature is a free-text description of the service provided by the supplier. The next two identify the business units involved in the customer/ supplier relationship. The next feature describes the measures used in the SLA, and the remainder describe the relative importance of each of the business drivers on a one to ten scale (ten indicating the most important).

8.5.2   System design

The measure selection system consists of four main components, as shown in Fig. 8.3.

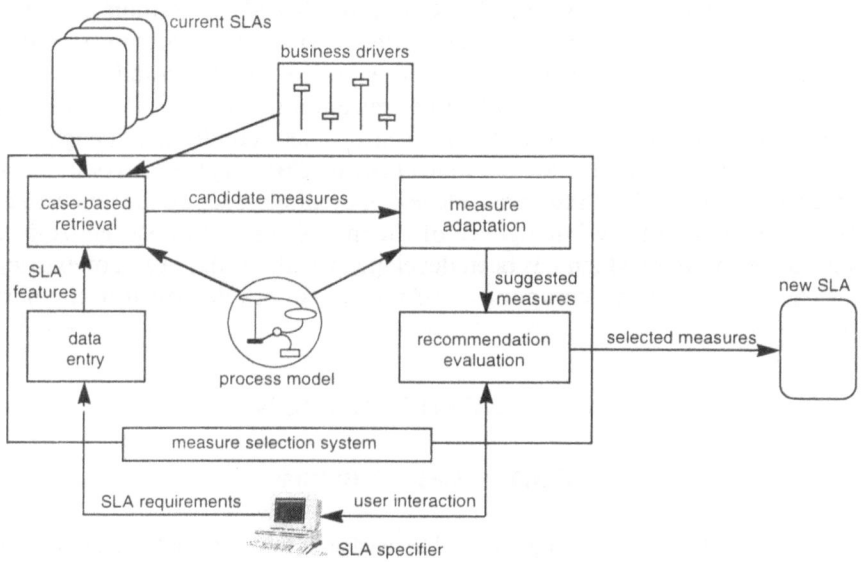

Fig. 8.3   Demonstrator functionality.

8.5.2.1   Entry of new situation data

The first component (data entry) is concerned with input of the problem and extraction of its important features.

Users are able to describe requirements for a new SLA using ideas and concepts with which they are familiar. They can select the interface between co-operating business units using a mouse pointer on a graphical representation of the business process. This action also identifies the customer, supplier and service description features of the new situation. The relative importance of the business drivers can be entered using a mouse-driven numeric value selector. Figure 8.4 shows a typical set of features for a repair SLA.

feature	value
service description	repair of payphones
customer	BT payphones
supplier	network maintenance
measures	?
importance of QoS	8
importance of cost	1
importance of quality of work	8
importance of efficiency	5
importance of revenue	3
importance of market share	5

Fig. 8.4   Features of an example new situation.

8.5.2.2   Retrieval of similar cases

The second component (case-based retrieval) is concerned with retrieval of relevant past solutions (i.e. SLAs). Twenty current BT SLAs have been stored as cases, an example of which is given in Fig. 8.5. The features of the new situation are compared with those of all the SLAs in the case base, and the cases which are most similar to the new situation are identified (e.g. Fig. 8.5 is a sufficiently close match to the new situation shown in Fig. 8.4 for it to be selected).

feature	value
service description	diagnosis of private circuit faults
customer	private circuit control
supplier	network maintenance
measures	measure 1 measure 2
importance of QoS	8
importance of cost	3
importance of quality of work	8
importance of efficiency	6
importance of revenue	3
importance of market share	5

Fig. 8.5   Features of an example case.

The usefulness of a previous solution depends on how similar it is to the new situation. Normally, the more similarities there are between two situations, the more likely it is that the solution used for one will also work for the other. In the simplest case, therefore, it should be possible to count how many of the SLA features are the same, and choose the case with most matches.

However, in most situations some features carry more weight than others. Where this is the case, the matching algorithm must take this into account, so that the best match is the one with most matches between the most important features.

In the demonstrator system a weighted matching algorithm is therefore used to give each case a score — up to a maximum of 100. After some experimentation, the weights for the features were chosen as follows:

- service description (30%) — a maximum score is given for a perfect match, less for a partial match;

- customer/supplier (5% each) — a maximum score is given if the feature matches, or zero if it does not;

- importance of business drivers (10% each) — the score depends upon the closeness of the values, a maximum score being given for a perfect match.

This algorithm is used as the basis for identifying the three SLAs with the highest score, and the measures used in their definition are passed to the third part of the system.

8.5.2.3   Measure adaptation

The third component (measure adaptation) is concerned with adapting the measures to the new situation, using them as the basis for defining a set of measures for the new SLA. It is this component which embodies much of the intelligence shown by CBR systems.

In order to achieve this, it has been necessary to construct a model of the measures used in SLAs throughout BT. This model is in the form of a type hierarchy — the measures becoming more specific, the lower they are in the hierarchy (see Fig. 8.6 for a partial view).

It has also been necessary to construct simple models of each of the interfaces between business units. Application of the hierarchy and these models allows the measures used in the chosen SLAs to be adapted to the new situation. This is achieved by identifying, for each of the selected measures, the generic measure type from which it was derived, then constructing an instance of it which is specific to the new interface.

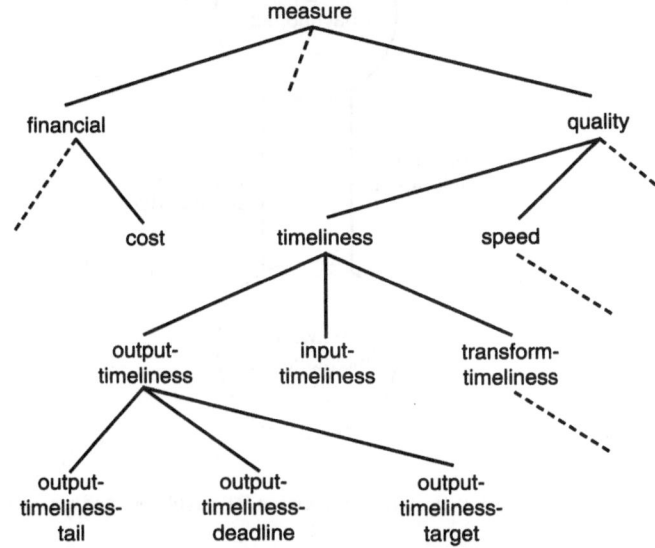

Fig. 8.6   Partial measure class hierarchy.

If, for example, measure 1 in Fig. 8.6 was 'the percentage of private circuit faults diagnosed within X hours of reception' (which is a measure of type 'output timeliness target'), this measure would be adapted using the model of the interface for the new situation (see Fig. 8.7).

The measure suggested for use in the new SLA would be 'the percentage of payphone fault reports cleared within X hours of reception'.

In the example above, the adaptation required is relatively straightforward. In other examples, where measures are taken from cases which do not match so closely, the required measure adaptation is more complex.

8.5.2.4   Evaluation of recommendations

The final component (recommendation evaluation) is concerned with evaluating the recommended set of SLA measures. In the demonstrator system, this is achieved manually, by presentation of the measures to the user of the system. The user can accept some or all of them, or suggest alternatives.

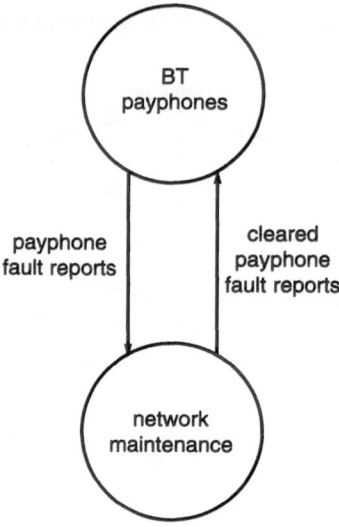

Fig. 8.7   Simplified model of the interface between BT payphones and network maintenance.

8.6   DISCUSSION

8.6.1   System performance

The case-based retrieval component identifies the most closely matching SLAs within a few seconds, and the measure adaptation component is equally fast in proposing a set of measures for the new situation. There is every reason to believe that good system performance can be maintained as the size of the case base increases.

The solutions proposed by the system for various business interfaces have been shown to experts, who have agreed that, in general, they represent sensible recommendations. However, when compared to the existing SLA for a particular interface, these solutions do not always correspond exactly. There are a number of reasons for this:

- some of the measures in the existing SLA are unknown to the CBR system;

- the case base is small (only 20 cases), so the best matches in the current case base may not be the best possible;

- the business drivers selected may not be exactly the same as those in force when the SLA was specified;

- the measures originally selected for the SLA may not have been most appropriate.

8.6.2   Benefits of using CBR

The implementation of the measure-selection system using CBR has allowed the following benefits of CBR to be demonstrated.

- Knowledge acquisition is relatively straightforward — it was not necessary to spend time identifying and interviewing experts, specifying and designing large amounts of software, or developing complex models. Instead, copies of existing SLAs were obtained, the key features and measures were extracted manually (approximately 20 minutes per SLA) and the details entered into the case base. This is a much simpler process than that used for more traditional information processing techniques.

- Past experience can be used to guide generation of solutions — historically, it has been very difficult for automated systems to represent experiential knowledge (i.e. knowledge gained by previous experience of performing a task), and to make it available to users. However, it is exactly this kind of knowledge that is normally used by skilled operators (who ask questions such as 'What caused the problem last time I saw these symptoms?' and, 'What did I do then to overcome it?'). The CBR system described in this chapter allows the experience of SLA specifiers (which is implicit in the measures they select for each SLA) to be captured and used.

- A solution can be produced where current practices are not clearly defined — developing a model of the measure-selection procedure would be very difficult using traditional techniques, because the process itself is poorly defined. Currently, the choice of measures is affected by a number of interrelated influences and is therefore based on the intuition of groups of individuals. However, a CBR approach has allowed a usable system to be produced without the need to develop a detailed model of the process. It has also allowed a repository of corporate SLA knowledge to be captured and made available to non-experts.

- Solutions are generated using a consistent approach to the problem — currently, because the measure-selection process is ill-defined and is carried out by independent groups of people, there are many inconsistencies in the use of measures. Thus, groups of individuals from different parts of the company may suggest very different measures for similar types of SLAs. One effect of this, is that more than one measure may be implemented to meet the same purpose. There is a significant overhead associated with measuring, recording, processing and maintaining each of these redundant measures. CBR enforces a consistent approach to measure selection, because the specification of each SLA is dealt with in the same way. Hence, CBR helps to minimize the overhead caused by unneccessary measures and supports uniformity in use of measures.

- Incomplete information can be dealt with in a controlled manner — it is possible to enter an incomplete set of details into the system, and it will still recommend a feasible set of measures. This is because, unlike many less-flexible technologies, a CBR system does not require a complete set of information in order to produce a solution. It will simply match to the previous experiences that most closely approximate the new problem.

CBR also brings a number of additional benefits, which were not demonstrated in this system. It is possible to add 'bad' SLAs to the case-base, in order to indicate potentially unfruitful solutions. It is also possible for a case-based system to improve its performance as it is used. This is achieved by automatically adding to the case base any new SLAs which incorporate additional user expertise (i.e. any new SLA for which the measures recommended by the system were modified by the SLA specifier prior to agreement).

8.6.3   Drawbacks of CBR

Case-based reasoning, however, has two major drawbacks, as detailed below.

- Solutions are based only on previous experience — this means that major changes to the domain cannot easily be incorporated into the system (i.e. if a major change to the process is implemented, the existing case base will contain no SLAs which take account of the change. Therefore, when a new SLA is required the system will still recommend a set of measures, but this will only be based on the SLAs in the existing case base, so will not take account of the change). In order to overcome this, it is necessary either to combine CBR with other types of reasoning (e.g. rule-based or model-based systems), or to 'hand-craft' example SLAs. The former possibility is currently being investigated.

- Solution accuracy depends on the quality of information in the case base — the quality of the solution generated by a CBR system is dependent on the quality of the solutions which are stored in the case base. It is, therefore, necessary to 'vet' cases before they are entered into the case base. This vetting procedure should ensure that cases added to the case base are correctly described, represent good solutions to the problem (or are labelled as being bad solutions), and cover any gaps in the domain knowledge.

8.6.4   Future applications

CBR offers potential in many other application areas, where it is possible to obtain past histories of problems and their solutions. It is currently in use within BT, providing assistance for operators of a help-desk facility. British Airways are experimenting with a prototype CBR system for in-flight fault diagnosis [6], and the applicability of CBR to the reuse of engineering designs is also being studied [7]. Other possible applications include decision support for scheduling, process control, software design, intelligent text retrieval and project planning.

8.7   CONCLUSIONS

The main strength of CBR is that it allows solutions which have been generated for previous problems (and which contain a great deal of useful information) to be used to help guide in the solution of new problems. This allows the experiences of separate groups of individuals to be shared between everyone involved in a particular task. It therefore makes expertise more widely available and encourages consistency in problem solving.

REFERENCES

1.   Ward K (Ed-in-chief): 'Telecommunications engineering: a structured information programme', Institution of British Telecommunications Engineers (1993).

2.   Shapiro S (Ed): 'Encyclopedia of artificial intelligence', 2nd Edition, 2 , Wiley Interscience (1992).

3.   Hammond K J: 'Case-based planning: a framework for planning from experience', Cognitive Science, 14 , pp 385-443 (1990).

4.   Turner R M: 'Organising and using schematic knowledge for medical diagnosis', in Proceedings of a Workshop on Case-based Reasoning, Clearwater Beach, Florida, pp 435-446 (May 1988).

5.   Mott S: 'Case-based reasoning: market applications and fit with other technologies', Expert Systems with Applications, 6 , pp 97-104 (1993).

6.   Magaldi R B: 'CBR for troubleshooting aircraft on the flightline', IEE Colloquium on case-based reasoning: prospects for applications, London, pp 6/1-6/9 (March 1994).

7.   Domeshek E A and Holodner J L: 'Toward a case-based aid for conceptual design', International Journal of Expert Systems, 4 , No 2, pp 201-220 (1991).

9

DECISION SUPPORT FOR SERVICE PROVISIONING

P D O'Brien and R G Davison

9.1   INTRODUCTION

One of the most important demands on service providers is responsiveness to fulfilling customer needs. Service providers face the challenge of delivering customized service solutions, competitive pricing, short delivery periods and a high quality of service [1]. A service provider must strive for optimum exploitation of resources (personnel and equipment). Such optimization in a multiprovider, multiple-service and multiple-network environment can only be achieved by powerful computer-based tools. The premise is that leading-edge technologies, such as knowledge-based systems and operations research can meet the challenges of service provisioning.

Service provisioning requires a combination of human judgement and computing power. Human judgement to select a solution that best meets the customer's requirements and the business objectives of the service provider. Computing power to deal with the large quantities of information needed to manage today's networks and services. Decision support systems (DSSs) couple the intellectual resources of individuals with the capabilities of computer-based support systems [2]. Project DESSERT [3] has developed three decision-support systems to assist with the process of service provisioning.

9.2   SERVICE LIFE CYCLES AND SERVICE PROVISIONING

Service provisioning can be best understood by considering a complete lifespan of an advanced communications service [4]. This contains three separate timelines or life cycles (see Fig. 9.1) — service-type creation life cycle, service provisioning life cycle and service usage life cycle. A life cycle provides a partially-ordered set of operations or functions:

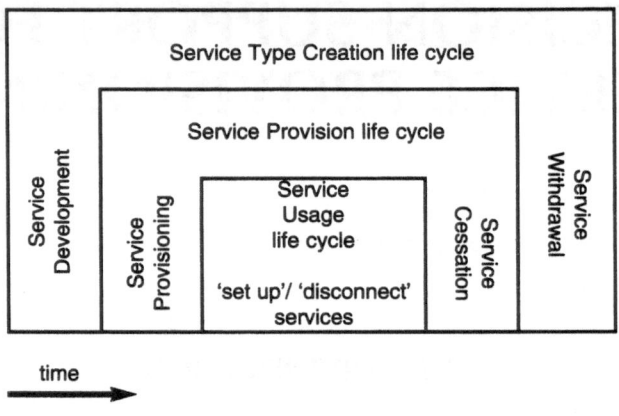

Fig. 9.1   Service life cycles.

- the service-type creation life cycle consists of those management activities involved in the construction, modification and withdrawal of a type of service;

- the service provisioning life cycle involves those activities necessary to create and delete an instance of a service for use by a particular customer — there are two phases: service provisioning and service cessation;

- the service usage life cycle involves those activities required to support a customer's use of a service which is part of the behaviour of that service — such behaviour may include the ability to set up and release calls and it may include the ability to customize the service to suit its users.

Service provisioning is responsible for capturing customer's requirements, identifying suitable service offerings, negotiating contracts, and allocating appropriate network, computing and installation resources to support a service offering. The initiator for service provisioning would be a particular customer expressing a need for a service. The initiator for service cessation would be either the customer (e.g. requesting discontinuation), the provider (e.g. failure to pay bill) or the contractor, (e.g. the duration of contract ends). Figure 9.2 illustrates the component management operations of the service provisioning life cycle. A fuller description of service provisioning and the service life cycles is given in Specification H.414 [5].

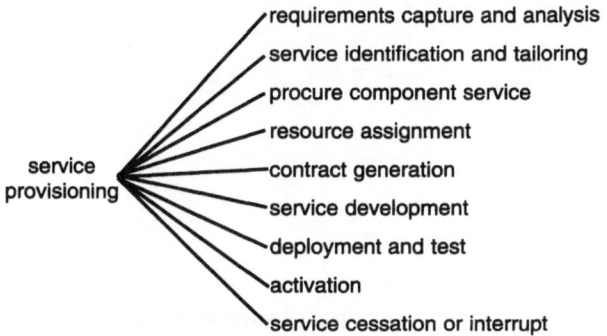

Fig. 9.2   Functional decomposition of service provisioning.

9.3   THE 'DESSERT' SCENARIO

Producing a successful DSS requires an understanding of the context in which it will be used. The DESSERT project has created a realistic scenario in which to test the DSS demonstrators. This is based on a vision of future services and networks in an open and competitive telecommunications market. The scenario involves three actors:

- the customer;
- the transport providers or long-distance carriers;
- the service provider (which the project has named 'Serv-U').

Their areas of responsibility are delineated by the shaded areas in Fig. 9.3. Serv-U re-sells communications services bought from transport providers, and combines them with its own computing and network resources to offer a variety of broadband services. Serv-U's service catalogue consists of a variety of broadband services, ranging from document handling through to videoconferencing services. The service catalogue is flexible, representing service types as a combination of component service elements.

In this scenario, Serv-U will use a number of DSS to determine how best to fulfil the customer's service requirements by optimizing the use of its own resources as well as its use of transport-providing services. At the same time Serv-U has to ensure an accurate and timely provision of the service to the customer.

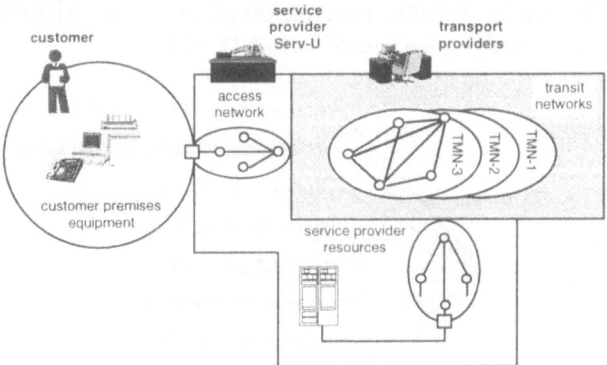

Fig. 9.3   The DESSERT scenario.

Serv-U owns communications resources forming an access network that is used to connect its customers to transport providers' networks. The access network is a mixture of ISDN (integrated services digital network), PPSDN (public packet-switched data network) and MAN (metropolitan area network) technologies. DESSERT DSSs are intended for use by Serv-U's personnel who are assumed to have expertise in the domain of service management and to be computer literate.

9.4   SERVICE PROVISIONING DSS DEMONSTRATOR

DESSERT has selected three areas of service provisioning and has developed a demonstrator DSS for each area, as follows:

- customer requirements capture DSS (CRC) — which supports its user in capturing a customer's requirements and converting them into a technical specification [6];

- generation and selection of alternative configuration DSS (GSAC) — which supports its user in taking a set of technical, network-independent specifications for required services and producing a network configuration;

- resource scheduling DSS (RSA) — which supports its user in scheduling Serv-U's workforce to implement a configuration [7].

In the following sections the GSAC demonstrator is chosen as the example described in detail.

9.4.1   GSAC — the system and user roles

The GSAC DSS supports its user in performing resource allocation within service provisioning. This is the process of identifying service and network resources required to generate a particular set of customer services. The process will consider the requirements of the customer and of the provider to ensure that the needs of both are met. It is likely that there will be more than one possible solution. The best solution will be chosen where 'best' is judged on a mixture of quantitative criteria such as installation cost and operation cost, and qualitative criteria such as security and extensibility. Performing resource allocation requires a mixture of activities, some of which are computationally intensive, such as optimizing bandwidth usage, and some of which are knowledge intensive, such as ensuring that a set of resources meets a security requirement.

In the design of decision support systems it is important to consider which aspects of the overall functionality will be performed by the user and which by the DSS. For the GSAC demonstrator the role of the DSS is as follows.

- To order the decision process — producing a resource allocation requires making a series of decisions. The order in which these decisions are made will affect the quality of the final solution and will differ between allocations. While people are able to identify a good ordering in isolated problems it is believed that they would need support in identifying a good ordering when dealing with several problems at once.

- To perform computationally intensive sub-decisions — certain sub-tasks of creating a configuration are best performed by the GSAC. This can be because the computational complexity involved in processing the task puts it beyond human capabilities or because, although the computation is simple, the quantity of data that needs to be considered results in greater efficiency if the task is automated. For instance, the optimal placing of required bandwidth to support services on to the available circuit bandwidth is a task best performed by the system because the complexity of the computation stretches human ability.

The user is required to perform three aspects that are not fully amenable to automation.

- Multi-criteria judgements with qualitative criteria — the selection of the best configuration is a multicriteria decision that includes quantitative criteria such as installation cost and running cost and less quantifiable criteria such as:

— the reliability of a configuration;

— the difficulty of maintaining a configuration;

— the extensibility of a configuration to meet future demand;

— marketing factors such as treating a particular service as a loss-leader.

Although GSAC can provide support as will be described later, only the user can decide how important the criteria are and make the final choice between configurations.

- Coping with impossible requirements — the requirements passed to the GSAC will always be technically achievable. However, it may be impossible to achieve them while satisfying the business requirements of the service provider. In this situation the user will be required to relax requirements so that a compromise can be reached.

- Coping with unforeseen circumstances — the GSAC designer is unlikely to have foreseen all sets of requirements. The experience and intuition of the user can be utilized to choose an approach for creating the configuration. This may be where the system is unable to cope or where the user is able to recognize a more efficient approach.

9.4.2   GSAC design

GSAC consists of a number of tools integrated via a blackboard mechanism [8], together with a control tool, named the 'Task Orderer'. Each tool either automates a specific service-provisioning task or supports the user in performing a specific service-provisioning task as shown in Fig. 9.4.

The blackboard mechanism is analogous to a group of human experts working together and using a blackboard to record their results. In its software implementation, the blackboard provides structured shared memory for problems and solutions, and provides a simple and flexible mechanism to pass data.

Each tool performs a specific task which could progress a problem towards resolution. Tools examine problems on the blackboard and notify a control tool (the Task Orderer in GSAC) when they can contribute. The control tool then decides which tool will work on each problem. Tools can compete with each other (i.e. a number of tools perform the same task but adopt a different approach to solving it), or co-operate with each other (i.e. a combination of tools work in series to resolve a problem). The user interacts with the individual tools to solve problems.

The blackboard approach has two key advantages for GSAC:

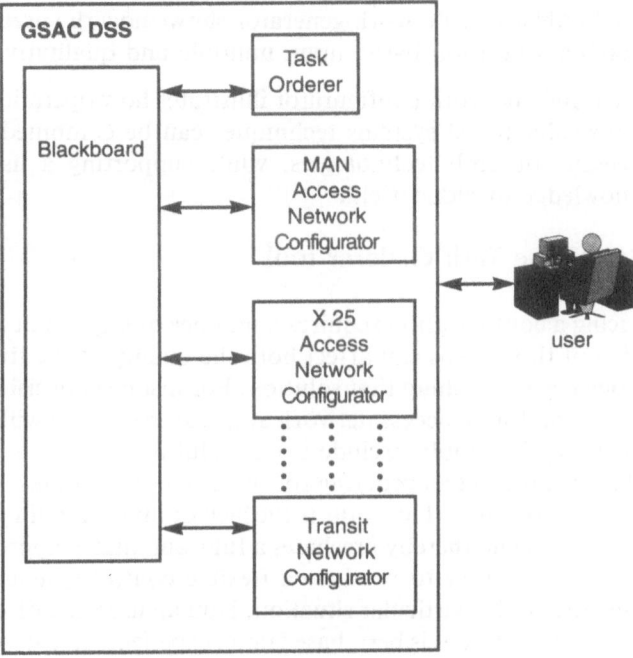

Fig. 9.4   GSAC design.

- each tool has to be connected to the blackboard but has no dependency on the other tools — therefore tools can be added or removed with the minimum of effort, resulting in a flexible and extensible framework for integrating a diverse range of tools and techniques;

- there is minimal predetermination of the order in which tools will be used preventing unnecessary restriction on the use of GSAC — other approaches such as procedural control require the system designer to encode all orderings envizaged as useful and to restrict the system to those orderings.

9.4.3   GSAC tools

Three example tools are described in this section each illustrating different aspects of decision-support systems:

- the Task Orderer is an example of how the problem solving control tool can provide a flexible and adaptive control mechanism for DSS;

- the MAN access network generator shows how the user is supported in making a decision based upon multiple and qualitative criteria,

- the transit network configurator illustrates how operations research and knowledge-based systems techniques can be combined to provide the benefits of both technologies, while supporting a user who has no knowledge of either field.

9.4.3.1 The Task Orderer tool

Producing a configuration requires a number of steps to be performed. The ordering of those steps can affect both the quality of the final solution and the efficiency of reaching that solution. For instance, decisions made on the configuration for an access network at a customer's site will restrict options elsewhere, and possibly exclude better solutions. To date, it has not been possible to construct an ordering for these steps that is best for a sufficiently high number of cases. This reflects the lack of understanding of the structure of the problem and thereby precludes a fully automated approach. As a result it has been necessary to produce a flexible control system that can adapt ordering to suit the particular situation. Human users are often able to decide which ordering of steps is best, based upon experience and domain knowledge, and therefore it is important for the user to have a major role in this process. However, the scale of the problem does mean that it can be difficult for users to keep track of what should be done and what has been done. To provide this support a tool called the Task Orderer has been produced.

The Task Orderer provides decision support to the user by continually monitoring the progress towards the problem solution, advising the user of which tasks could be carried out next and recommending which of those tasks should be addressed. The user can then select which task to perform based on the Task Orderer's recommendations. Raghavan [9] uses the term 'active decision support' to describe systems that provide autonomous support to a user. The Task Orderer falls into this category.

The user is presented with two lists (see Fig. 9.5) — one showing the tasks currently being undertaken and one showing the tasks that could also be undertaken. If the user wishes to undertake more tasks they can select, by using this window, which one to start. The list of possible next tasks is ordered by the Task Orderer to provide a recommendation to the user of what should be attempted next.

The Task Orderer has been implemented by extending the agenda-based control mechanism adopted in most blackboard systems. In a traditional blackboard system [8], whenever a tool can perform some work it informs a blackboard controller, which obeys some predefined strategy in determining

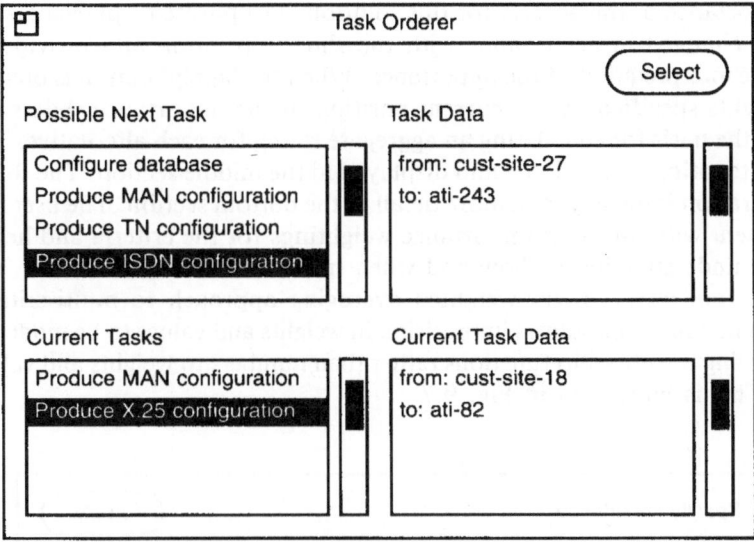

Fig. 9.5   A typical Task Orderer screen.

which tool to execute. In GSAC this has been extended to involve the user, so that when tools can perform they inform the Task Orderer, which presents the user with a prioritized list of tools that can then be activated.

9.4.3.2   The MAN Access Network Configuration tool

The MAN-ANC tool supports the user in the generation of an access network configuration that will connect the customer to the transit network with the required quality of service. It provides an illustration of how advanced information processing techniques can support decision making based upon multiple and qualitative criteria.

The basic operation of the tool is to generate the technically feasible alternatives and to evaluate which best meets both the customer's and the provider's requirements. The alternatives are generated by using a rule-base of design knowledge and are then passed to the MAN configuration evaluator.

The MAN configuration evaluator assists the user in understanding the trade-offs that exist between possible solutions. For example, it may be that the most reliable solution is the least extensible. It also supports the user in giving different importance weightings to criteria and selecting the best solution. An example is illustrated in Fig. 9.6. The top section contains a number of criteria which are the relevant factors used to judge a configuration. It provides a set of sliders which the user moves to specify the importance of the criteria for this solution. The phrase displayed on the right is changed in accordance with the slider's position and provides a linguistic interpretation of the importance. When the 'apply' button is pressed the weights specified by the current position of the importance sliders are used as the basis for calculating an aggregate rating for each alternative. The aggregate ratings are ordered and displayed in the middle section. The 'best' configuration is displayed in more detail in the bottom section. The user can experiment with different importance weightings for the criteria and hence gain an understanding of how and when trade-offs occur.

This tool uses a fuzzy-weighted averaging approach to multi-criteria decision making. This allows imprecision in weights and values to be modelled by assigning membership functions rather than numbers to weights and values [10]. This is illustrated in Fig. 9.7.

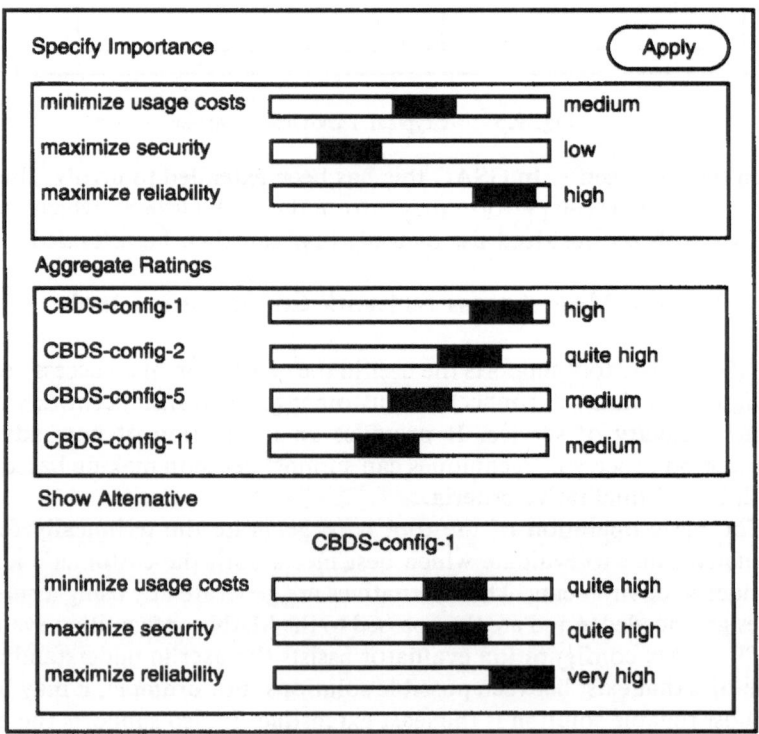

Fig. 9.6   MAN configuration evaluator.

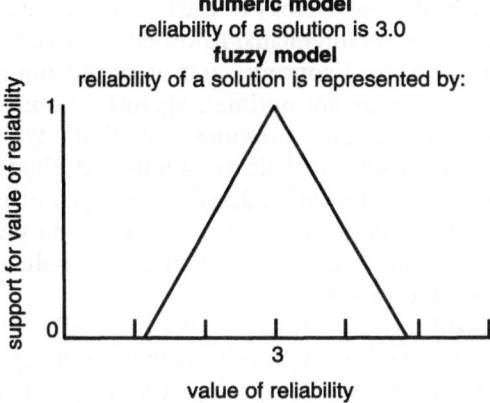

Fig. 9.7   Fuzzy modelling versus numeric modelling.

In the MAN configuration evaluation tool, each configuration is assigned a fuzzy value for each of the criteria. The importance given by the user to each criterion provides fuzzy weights that are used to produce the weighted average for each configuration.

Further information on the use of fuzzy-weighted averaging in multi-criteria decision making is given in Brown [11].

9.4.3.3   Transit network configuration

The transit network configurator (TNC) tool supports the user in placing a set of specifications corresponding to customers' requirements on the bandwidth available across a transit network. It provides an example of how operations research (OR) and knowledge-based systems (KBS) techniques can be combined to good effect.

In the DESSERT scenario, the transit network can be viewed as a collection of large bandwidth circuits, each with defined cost and quality, and capable of supporting a number of customer requirements. The user and the tool configure the network by placing the customer requirements on those circuits. The specification of a customer's requirements for the transit network will describe required bandwidth, usage information and the required quality of service which may include factors such as security, reliability and availability. The aim of the transit network configuration tool is to meet those requirements to a level acceptable to the customer while meeting the business objectives of the service provider.

Operations research techniques have been used to ensure bandwidth requirements are met while minimizing a single cost function such as price.

However, the best configuration may be the one that is cheapest, most reliable and best meets the usage requirements. Knowledge-based techniques are good at handling such ill-defined concepts and at producing solutions that are acceptable, though perhaps not optimal, against a number of criteria. The transit network configurator provides functionality based on linear programming and on KBS heuristic techniques. It chooses between these techniques based on a model which describes the type of problem that each is able to solve best. The user is asked to set some parameters to help characterize the problem. These are all related to domain aspects and do not require any knowledge of OR or KBS.

Once a solution has been produced, the tool will critique the solution, judging how well it meets the input criteria and reporting its opinion to the user. This is achieved by comparing the original quality-of-service requirements input to the tool with the quality-of-service exhibited by the solution. Any differences are reported to the user, who, if dissatisfied, can alter the matching accuracy that is attached to each criteria and produce a different solution.

9.5   THE 'DESSERT' TOOLKIT CONCEPT

The production of service management DSS is an expensive process requiring highly skilled personnel. Investigation of the service management domain has identified numerous similar tasks that present an opportunity for software reuse. An architecture has been produced to structure DSS software in a manner amenable to reuse at both a domain and technology level. Project DESSERT has implemented aspects of this architecture, and developed a DSS toolkit and methodology. The DSS toolkit supports the production of service management DSS by taking an 'assembly-by-parts' approach of 'plugging' together reusable components to form a new DSS.

According to Leitch and Stefanini [12], a toolkit provides a set of representations and control mechanisms restricted to a particular application domain, and additionally provides domain-dependent support to aid a system developer. The main advantage of such a toolkit over an expert system shell or high-level programming language is the degree of support provided to the toolkit user. A toolkit is targeted at a specific application domain, and as such can incorporate a large degree of domain-specific knowledge. As a result, the range of the system is limited, but the power or support given to the toolkit user is considerable.

The DESSERT toolkit architecture provides a framework for the managed reuse of software at both the task and the technology levels. There is a clear separation of service management domain tasks from the underlying software functionality that supports them. This has allowed the work on the DSS demonstrators to identify service management tasks as reusable primitives for building applications. Also, it has provided a structure for OR and KBS techniques to be reused within different tools.

It is anticipated that such a toolkit approach will have a number of advantages in that it will:

- support the rapid and efficient development of service management DSS;

- enable the reuse of software;

- reduce development costs for the service provider;

- simplify the development of service management systems by empowering domain experts to do the task.

Further information on the development of a DSS toolkit is given in Weigand et al [13].

9.6   CONCLUSIONS

Managing the problems associated with a growing portfolio of services will be one of the greatest challenges for service providers in the future. Supporting personnel in handling this task is vital to maintaining a competitive edge in future telecommunications markets.

Research has demonstrated how decision support systems and leading edge technologies can be used in the automation of service provisioning. It has illustated, through its DSS demonstrators, how a partnership of user and computer can effectively deal with the types of decisions inherent in the service management domain. A combination of knowledge-based systems and operations research techniques have been effectively applied in the development of such systems.

The use of DSS in provisioning will ensure that there is consistency of price, design and service across a service provider organization. Furthermore, the power of DSS will enable a user to deal effectively with the increasing complexity and volume of information used during service provisioning.

REFERENCES

1.   Basseil R and Gotz B: 'Integrated service management for intelligent networks', XII International Switching Symposium, Stockholm (1990).

2. Keen P and Scott-Morton M: 'DSS: an organisational perspective', Addison Wesley (1978).

3. 'DESSERT (Project R2021)', collaborative project, jointly sponsored by the CEC and contributing partner organizations via the RACE II programme, involving Broadcom Eireann, Framentec-Cogitech, SEMA Telecom, PTT Netherlands Research, Inform, Trinity College Dublin, Dublin City University, Queen Mary and Westfield College, BT (1994).

4. Davison R and O'Brien P: 'Service provisioning in a multi-provider environment', in Proceedings of the Intelligent Services and Networks Conference, Springer Verlag, Lecture Notes in Computer Science Series (1994).

5. 'Customer service provisioning: common functional specification H.414', RACE Central Office, Brussels (1994).

6. Tattersall C, Groote J and Pazuelo G: 'Satisfying enterprise-wide telecommunications needs: decision support for requirements engineering', in Spaniol O, Bauerfeld W and Williams F (Eds): 'Proceedings of the 3rd Broadband Island Conference', Elsevier (1994).

7. Dexheimer O, King S, Petit-jean S and Porte N: 'A software architecture for scheduling service and network configuration installation', ISN Conference (1992).

8. Nii H P: 'Blackboard systems (part one): the blackboard model of problem solving and the evolution of blackboard architectures', The AI Magazine, pp 38-53 (1986).

9. Raghavan S A J: 'A paradigm for active decision support', Decision support systems, 7, pp 379-395 (1991).

10. Zadeh L A: 'Fuzzy sets', Information and Control, 8 pp 330-353 (1965).

11. Brown T: 'A new fuzzy weighted average algorithm', in Piera Cerrete N and Singh M G (Eds): 'Proceedings of the IMACS International Workshop on Qualitative Reasoning and Decision Technologies QUARDET '93', CIMNE, Barcelona (1993).

12. Leitch R R and Stefanini A: 'QUIC: a development environment for knowledge based systems, in industrial automation', Proceedings of 3rd Esprit Technical Conference, pp 674-696, North-Holland (1988).

13. Weigand M, Davison R, O'Brien P, Reeder A and Alexander D: 'An architecture for developing DSS for service management', Proceedings of International Conference on Intelligence in Broadband Services and Networks (1993).

BIBLIOGRAPHY

Davison R, O'Brien P and O'Sullivan D: 'Decision support for configuring telecommunication services', Information and Decision Technologies, 19 , North Holland (1994).

O'Brien P, Davison R, O'Sullivan D, Parsons S and Mamdani E: 'DESSERT: decision support systems for provisioning telecom services over multiple networks', Proceedings of the Vienna IT Conference (1993).

Simon H A: 'New science of management decisions', Prentice Hall, New Jersey (1977).

Smith R, Mamdani E and Callaghan J (Eds): 'The management of telecommunications networks', Ellis Horwood, Chichester (1992).

Davison R, O'Brien P, Brown T and O'Sullivan D: 'A decision support system for the allocation of resources during service provisioning', Proceedings of the International Conference on Intelligence in Broadband Services and Networks (1993).

Commission of the European Communities: 'Perspectives for Advanced Communication in Europe, Impact Assessment and Forecasts', Issue 2, CEC, Office for Official Publications of the European Communities, Luxembourg (1992).

Sprague R H Jr: 'A framework for the development of decision support systems', in Sprague R H Jr and Watson H J (Eds): 'Decision support systems: putting theory into practice', Prentice-Hall, New Jersey (1986).

10

A GRAPHICAL TOOL FOR BILL DESIGN

P C Utton and P M Sanders

10.1   INTRODUCTION

The telephone bill is BT's most common form of communication with its customers. Over 100 million bills are sent out each year to more than 23 million customers. It is vital that these bills are easy to read and understand, and that they also convey a clear image of the company to its customers.

This chapter describes the design and implementation of SCRIBE (screen interface for bill editing), a tool to support the bill design process. In particular it is intended to assist with the redesign of bills, for example to improve clarity and information content, or to introduce billing categories for new services.

SCRIBE has been designed to provide a graphical user interface to BT's customer services system (CSS) with the following key features:

- accurate visual representation of bills on screen;

- 'what you see is what you get' (WYSIWYG) operation;

- advice to users on conformance to established design guidelines and system constraints;

- automatic generation of the formatting parameters required to drive the operational system.

Using SCRIBE, bills appear on screen exactly as they would on paper. This makes the consequences of any formatting decisions instantly apparent and enables the user to determine the most effective presentation for billing information. In addition, the WYSIWYG mode of operation simplifies changes to bill formats in that the user can 'grab' a graphical component on the screen, move it to a new position or change its contents, size and font, etc, without having to be concerned with its internal representation in terms of formatting parameters. Once an edit session is completed, the tool can translate the graphical representation of the bill into the information required by the operational billing system and download the formatting parameters to implement the new design.

Design may be viewed as a decision-making process. Within the context of this application, decisions are required on the effective presentation of information which can be readily achieved within the operational system. To this end, SCRIBE incorporates knowledge of various typographic guidelines and constraints imposed by the billing system. It uses a rule-base to evaluate layouts and provide advice to users. In particular, it attempts to provide early feedback on layout changes that cannot be supported by the main billing system. Collectively these facilities allow SCRIBE to function as a decision-support tool for the bill design process.

SCRIBE relates to the development and selection phases of the decision-process model introduced in Chapter 1. Design is largely an unstructured problem, which some would characterise as ill-defined [1]. SCRIBE's purpose is therefore to assist the user develop and evaluate alternate solutions, supporting a 'what-if' approach to problem solving. Its role is to support and not replace human decision making, and, in the final analysis, it relies on the user's judgement as to the aesthetic appeal of a layout.

The result is that this tool allows BT to be more responsive to customers by reducing the time to make changes to the billing process. It also provides a starting point for the introduction of more flexibility into the billing system, for example, in the form of customized bill presentation for selected customers. Very little information is available in the public domain about the billing support systems used in other telecommunications companies. However, it would appear that there are few companies which offer bill customization facilities at all.

The remainder of this chapter is structured as follows:

- section 10.2 provides background information on the SCRIBE development;

- section 10.3 discusses the design of the SCRIBE graphical user interface;

- section 10.4 covers the design of the constraint and rule-handling facilities;

- section 10.5 provides an overview of the implementation techniques used and, in particular, reviews the benefits that object orientation, functional languages and rule-based programming have brought to this application;

- section 10.6 contains the concluding remarks.

10.2 BACKGROUND

BT completely re-styled its bills in December 1992. This required major changes to the computer system that produced them — CSS [2]. At that time, the logic of the bill formatting process was modified to make it 'table driven'. The intention was that the bulk of future changes to bill layouts would then be achieved by changing data entries in a set of centrally held tables rather than requiring software code changes to CSS itself.

The end result was TABLET (Telecom's advanced bill layout and exchange tables) — a set of tables which define the content, position and presentation style of all the text items that appear on a bill, as well as various other rules about how bills are constructed. The scale of the billing operation, and the diversity of the products, services and billing categories that are supported, mean that the resulting tables contain over 6000 entries. It is therefore not surprising that format changes are still not as easy as one would wish.

The next logical step in the evolution of the billing system was to improve the user interface to TABLET. SCRIBE provides this new interface. It acts as a separate, self-contained 'front end' to TABLET. This strategy has been adopted for three reasons.

- It removes the constraints imposed by the TABLET implementation technology. TABLET runs in an IBM mainframe environment under the MVS^{TM}/CICS^{TM} operating systems and drives character-based terminals. It is written in COBOL and uses an IDMS^{TM} database. SCRIBE runs on a UNIX^{TM} platform under X Windows^{TM}, supports bit-mapped screens and provides an open systems solution.

- It provides a low-risk strategy for evolving one of the major systems. Such systems are often referred to as legacy systems — they represent a huge investment and provide the cornerstone of the operational infrastructure, but are typically extremely large and difficult to change.

- It means that much of SCRIBE is potentially reusable even if TABLET is replaced. In principle, a different back-end system could be accommodated by changing the data-conversion routines and supplying a different set of layout rules and constraints.

Figure 10.1 illustrates the overall context within which SCRIBE is intended to operate.

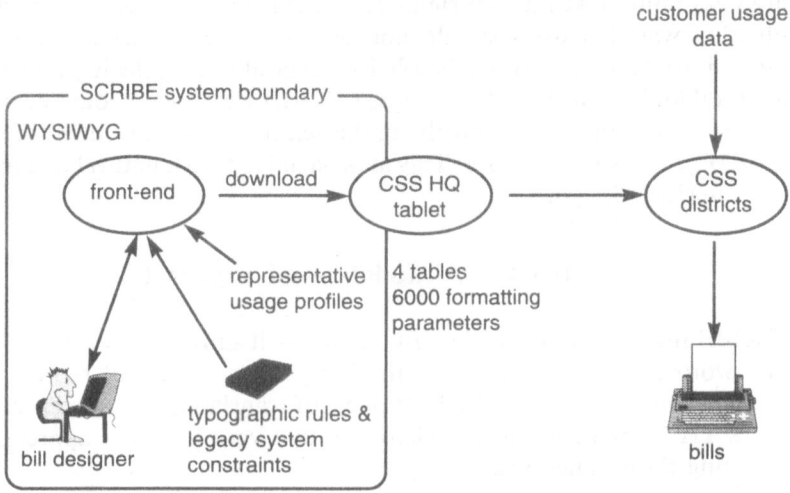

Fig. 10.1   SCRIBE — a front-end to TABLET.

10.3   THE 'SCRIBE' GRAPHICAL USER INTERFACE

The design of SCRIBE was complicated by the need to support the dynamic nature of bill formatting and the use of proportional fonts. Each bill is highly dependent on the customer's usage data and its format varies dynamically in two ways. Firstly, a billing category will only appear if it is relevant to the individual customer concerned, for example, references to a service that has not been used will be omitted rather than printed as a zero value entry. Secondly, and more importantly, the level of detail printed varies depending on how much space is available for printing. The first page of a bill, the summary page, is a legal document for the purposes of Value Added Tax, and must be printed on a single sheet of paper. If there is insufficient space to print the full details for a particular billing category then lower-level detail is 'compounded' (rolled up) into higher level categories on the summary page and explained in detail on the breakdown pages.

These formatting rules optimize the balance between information content and clarity for each bill. The use of proportional fonts also improves clarity but means that different characters in the same font, and the same character in different fonts, can occupy varying amounts of space on screen. The need to provide an accurate representation of a formatted bill on screen means that SCRIBE requires a sophisticated graphics capability with 'pixel level' control.

The desire to hide the complexity of the 'back-end' systems (i.e. TABLET and CSS billing) was also a significant influence on SCRIBE design. The intention was that users should not need to concern themselves with the internals of TABLET. A user's sole focus should be on the logical structure and total look of the bill. All changes to TABLET values should be inferred from graphical operations involving the selection and movement of screen-based objects, using a mouse. This approach leads to a natural and intuitive user interface design.

10.3.1   A model-based approach

SCRIBE must provide two key functions — it must act as a WYSIWYG-style word processor to produce the image of a document representing a telephone bill, and it must also be capable of translating this document image into a TABLET coded form. There are two very different approaches to providing these functions:

- a document-based approach, involving the use of an existing word processor and 'compilation' of its output into TABLET form;

- a model-based approach, where a model of a bill is constructed using an appropriate application-modelling tool-kit, which is then displayed to the user in bill image form and output to TABLET in an appropriately coded form.

The model-based approach was adopted as it was considered to offer greater flexibility and control in the design and operation of the tool. In fact, SCRIBE provides a number of different models or views of a telephone bill. Each of these models operates at a different level of abstraction, and is, deliberately, a simplification of the real-world billing system. Figure 10.2 illustrates the various models involved. The central SCRIBE composite model unifies the component models and provides for mapping between the SCRIBE and TABLET models.

10.3.2   The logical view

The logical view shows the structure of a bill in the form of a hierarchy (tree structure). This starts from a root node which corresponds to the entire bill, progresses via intermediate nodes which represent items such as summary and breakdown pages, main sections and sub-sections, etc, all the way down

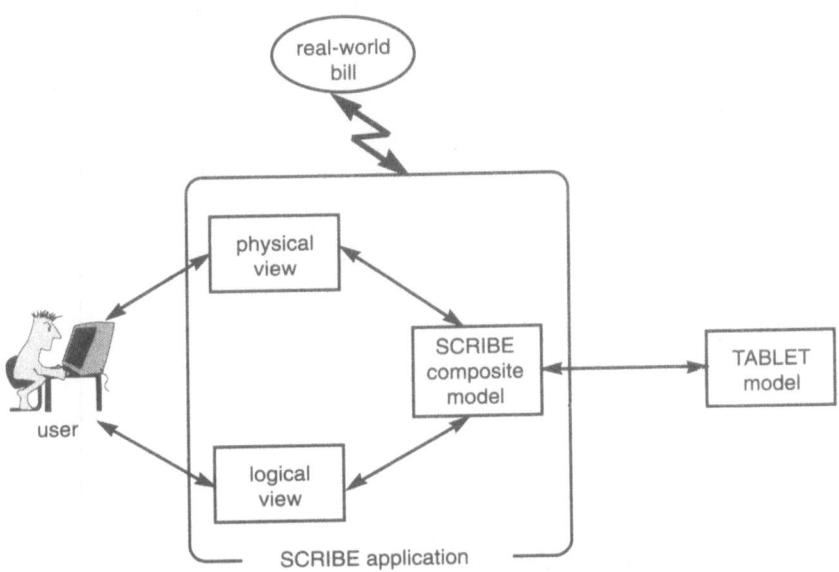

Fig. 10.2   SCRIBE model correspondence.

to the lowest, leaf-level entries which correspond to items such as itemized calls on a breakdown page (see Fig. 10.3 for an example logical view).

Operations within the logical view correspond to those of a 'tree editor', i.e. nodes within the tree structure may be created, modified and deleted. Bill items can be simply reordered by selecting and moving them on the screen, and entire substructures can be cut out or copied and pasted back into new positions to effect larger scale changes. Lower levels of detail can be temporarily 'folded' out of sight to reduce clutter on the screen.

10.3.3   The physical view

Physical details, such as position, font, etc, do not appear in the logical view. These aspects are covered by the physical view which provides a fully formatted image of the bill based on the example billing data in use. Example data values can be entered by overtyping in the appropriate regions of the bill (see Fig. 10.4 for an example physical view). The user can exercise a layout against different permutations of data by simply selecting different user profiles from a menu.

Example data may be loaded and saved separately from format data to facilitate reuse of different test cases.

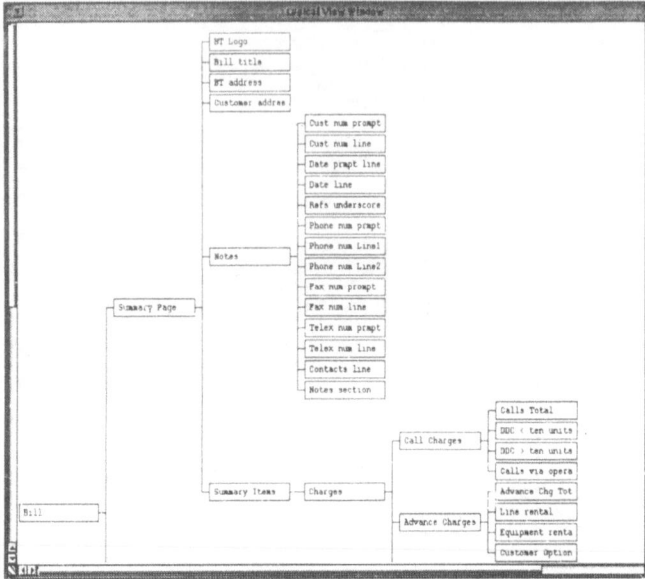

Fig. 10.3   A logical view window showing part of the bill structure (the remainder may be viewed by scrolling the display).

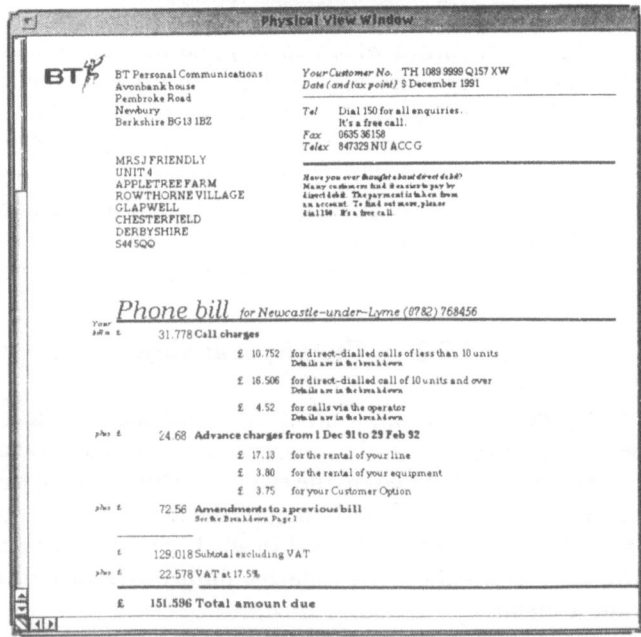

Fig. 10.4   A physical view window.

Each logical construct within the logical view is composed of one or more physical parts which appear within the physical view. Each physical part may have a different presentation format and will have a set of contents which may be one or more items. Physical parts need not be only text strings, but can also be lines (solid or dotted and of varying thickness), or indeed bit maps representing arbitrary graphical objects (e.g. a logo). Grabbing and moving an item in the physical view will cause its positioning information to be updated. The contents of any item may be changed by overtyping and the font size, type and style can be modified using menus of available options.

The physical view allows the user to speedily see the full impact of any changes in formatting rules and examine how these rules interact with different billing data. For example, entering a single line of new example data could have a dramatic effect on the overall presentation of the bill if it triggers the compounding process.

An advantage of offering multiple views within the user interface is that it provides different contexts and therefore possible interpretations for similar graphical operations. For example, grabbing a heading and moving it below another heading is an ambiguous operation. It could indicate that the headings are to be printed in a different order or it could indicate that ordering is to be preserved but that a larger gap should be left before the first heading. SCRIBE takes the former interpretation in the logical view and the latter interpretation in the physical view.

In addition, multiple-connected views help users navigate within the tool. For example, the logical view offers a route map into the physical view. Clicking on a logical construct will cause a physical view to be displayed which is centred on the matching physical parts. As an aid to ergonomics, it is also possible for the user to select a physical part within the physical view and be switched to a display centred on the owning logical construct within the logical view.

It has also been found beneficial for the tool to present a physical view on start-up which is pre-loaded with an existing format, as the emphasis when using the tool is on redesign rather than designing from scratch. Simple considerations such as these can have a significant impact on users' perceptions of the tool's usability.

10.4 'SCRIBE' CONSTRAINT AND RULE HANDLING

The discussion so far has focused on SCRIBE's graphical user interface. The other key requirement for the tool was to provide a layout evaluation capability that could guide users in the design of new layouts.

TABLET does not control every aspect of bill presentation. Some aspects are 'hard coded' into the billing system and some are essentially fixed by the markings of the pre-printed stationery. However, it is hard to decide by inspection of a printed bill which aspects are easy to change and which are difficult. SCRIBE advises on this type of issue. Its knowledge base covers:

- layout guidelines derived from typographic specifications;
- formatting rules which are embedded in the billing system;
- constraints reflecting limitations imposed by the operational system;
- other rules, guidelines and constraints identified by bill designers or the CSS billing team.

In the above, three different terms have been used for bill formatting knowledge — rule, guideline and constraint. In the real world, the implication is that a rule is something one must do, a guideline is something one should do and a constraint is something one must not do. To unify these concepts, from now on all of these categories of knowledge will be referred to as billing **directives** and directives can be advisory or mandatory[1].

To provide an effective design environment, it was considered important that the user should have the freedom to change any aspect of the look of the bill, but then be able to check for compliance with recommended practice. The danger was that 'locking' directives into the design environment would stifle designers' innovation. The evaluation function is therefore normally invoked at the user's discretion and will check a candidate design against the knowledge base and advise the user accordingly[2]. If a conflict with the directives is detected, feedback to users needs to clearly explain what the problem is and why it is considered a problem. The expressive power of the knowledge base is therefore important. SCRIBE should communicate with users in their own language to explain its advice. Feedback messages should be as explicit as possible.

Directives are likely to change over time as the billing system evolves. Furthermore, programming of the directives within SCRIBE might need to be modified quite urgently if errors were detected. (This is most likely to occur through imperfect knowledge of the behaviour of the billing system.) The key requirement is for the knowledge base to be easy to change and extend.

---

[1] The term **directive** is also used to distinguish this real-world knowledge from possible implementation constructs such as rule and constraint that might be used to represent the directives within the knowledge base. However, constraints, in the sense of constraint logic programming, have not been used in the SCRIBE implementation.

[2] Electing to download formatting parameters to the back-end system will also automatically invoke evaluation of the bill layout to ensure the revised format is supportable.

Figure 10.5 illustrates some evaluation output. It is structured in terms of what the problem is, why the tool considers it a problem and a choice of actions that the user can take. The objects referred to in the messages would be highlighted in the physical view on screen.

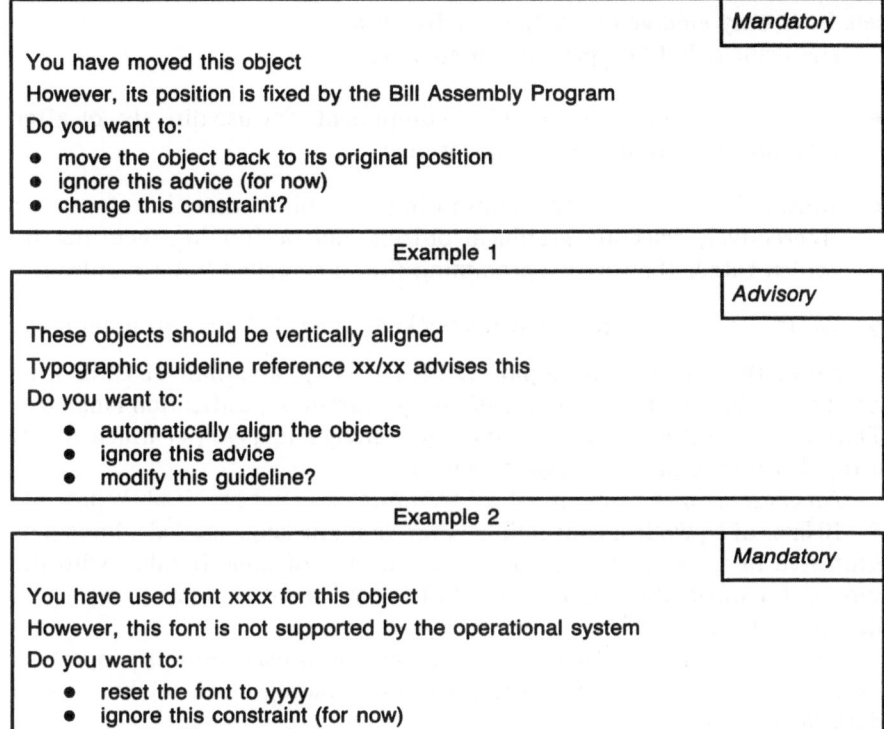

Mandatory

You have moved this object
However, its position is fixed by the Bill Assembly Program
Do you want to:
- move the object back to its original position
- ignore this advice (for now)
- change this constraint?

Example 1

Advisory

These objects should be vertically aligned
Typographic guideline reference xx/xx advises this
Do you want to:
- automatically align the objects
- ignore this advice
- modify this guideline?

Example 2

Mandatory

You have used font xxxx for this object
However, this font is not supported by the operational system
Do you want to:
- reset the font to yyyy
- ignore this constraint (for now)
- modify this constraint?

Example 3

Fig 10.5   Illustrations of layout evaluation feedback.

The user's response to evaluation feedback will normally be one of three options — either to accept the directive and so modify the layout, to disagree with it and attempt to change it, or simply to ignore it for the time being. Electing to ignore a directive will inhibit its further evaluation on the particular objects in question for the remainder of the session. Electing to change a directive will, assuming the user has the appropriate privileges, cause further dialogue boxes to be displayed to clarify the change to be made. It will also log the modifications to provide an audit trail. Accepting a directive will cause SCRIBE to exploit certain 'prototypical' information stored for each billing construct, and 'repair' the layout automatically.

10.5   IMPLEMENTATION

SCRIBE is implemented using a suite of tools supplied by ILOG Ltd[1]. It is probably true to say that the implementation of SCRIBE would not have been practical without the use of a modern integrated tool-set. Such tool-sets have only emerged over the last five years.

The principal ILOG products used were:

- Aida[TM] — a library of graphical components for use directly, or after adaptation, within users' applications;

- Masai[TM] — a tool for composing graphical user interfaces by interactively selecting graphical objects and positioning them on the screen rather than by programming (i.e. a so-called 'GUI builder');

- SMECI[TM] — an expert systems package for rule-based programming.

All of the ILOG tools employ LeLISP[TM] [3], a variant of LISP with an object oriented extension, as their programming/specification language. The core of LeLISP is a functional programming language [4] which allows a rapid prototyping development scheme.

An expression in a functional programming language typically represents 5—10 lines of equivalent conventional programming language code. Functions expressed in LISP tend to be succinct, and, to someone familiar with the syntax, far more understandable than the equivalent statements expressed in a more traditional procedural language, such as C.

LISP is designed to facilitate the processing of lists and this property is exploited within the SCRIBE application by holding much of the internal data in this form.

Much of the power of the chosen tool-set comes from its support for the object oriented model. In particular **encapsulation** of code and data within objects and **inheritance** of properties via class hierarchies are key beneficial features. These facilitate the adaptation of components from the Aida graphics library. Cusack and Cordingley [5] provide a simple introduction to object oriented concepts; Booch [6] and Meyer [7] are more comprehensive texts.

The combination of functional programming in LeLISP and object oriented data structures in MicroCeyx greatly simplifies the programming task. As an indication of this, only 1400 lines of application codes are required to implement the SCRIBE 'display engine' which provides the logical and physical views and the simulation of the CSS bill-assembly process.

---

[1] ILOG was set up to develop commercially ideas originating within the French software research establishment, INRIA.

SCRIBE has been designed so that its graphical design capability is decoupled from its layout evaluation capability. Its knowledge base is a distinct component from its display engine and is implemented using SMECI [8-10]. SMECI is a development environment offering object-based knowledge representation and an inferencing capability using 'production' rules. The key feature of these rules is that they present premise/conclusion pairs, i.e. if a certain premise is true, it may be deduced that a certain conclusion is true. Winston [11] provides a readable introduction to this field.

All forms of billing directive (i.e. rules, guidelines and constraints) are represented by SMECI rules. The benefits provided by the SMECI approach are primarily the declarative style of the rules, their user-friendly language, and the fact that they are readily inspectable. In addition it is possible to add new rules to an operational system incrementally.

10.6   CONCLUSIONS

This chapter has outlined the design philosophy behind a new graphical front end to BT's telephone billing system. To summarize, the design imperatives were to:

- communicate meaningfully with the user, in a natural and intuitive fashion;

- hide the complexities of the back-end billing process;

- map to TABLET;

- be flexible.

SCRIBE provides decision support by enabling users to quickly decide how they would like to present billing information and then to check how achievable their ideas are in practice. Although the emphasis is on guiding the user to operate within the directives imposed by the operational system, SCRIBE also promises to provide information on whether particular directives might really need to be modified.

APPENDIX

Glossary

CICS   A transaction processing monitor for IBM mainframes
CSS   BT's customer services system

GUI   Graphical user interface
IDMS   A mainframe database management system supporting the
              network model of data
MVS   An IBM mainframe operating system
SCRIBE   Screen interface for bill editing
SMECI   An expert system package supplied by ILOG Ltd
TABLET   Advanced bill layout and exchange tables
WYSIWYG   'What you see is what you get'

REFERENCES

1. Simon H A: 'The structure of ill-structured problems', Artificial Intelligence, 4 , pp 181-200 (1973).

2. Rockliffe R: 'Implementing BT's New Bill', British Telecommunications Eng J', 11 , pp 273-278 (January 1993).

3. Chailloux J et al: 'Le-LISP: a portable and efficient LISP system', Proc ACM Symp on Lisp and Functional Programming, ACM Press (1984).

4. Hudak P: 'Conception, evolution and application of functional programming languages', Computer Surveys', 21 , No 3, pp 359-411 (1989).

5. Cusack E L and Cordingley E S: 'Object oriented techniques in tele-communications', BT Telecommunications Series, Chapman & Hall (1995).

6. Booch G: 'Object oriented design with applications', Benjamin/Cummings Publishing Co (1991).

7. Meyer B: 'Object oriented software construction', Prentice-Hall International Series in Computer Science (1988).

8. Neveu B and Haren P: 'SMECI: an expert system shell for civil engineering', Applications of Artificial Intelligence in Engineering Problems, First International Conference, Southampton University, Springer-Verlag, New York (April 1986).

9. Neveu B, Trousse B and Corby O: 'SMECI: an expert system shell that fits engineering design', 3rd International Symposium on Artificial Intelligence, ITESM, Monterey, Mexico (October 1990).

10. Dieng R, Corby O and Haren P: 'Explanatory knowledge tools for expert systems', 2nd International Conference on Applications of Artificial Intelligence, Boston MA (1989).

11. Winston P H: 'Artificial intelligence', Addison-Wesley Publishing Company (1979).

11

PLANNING BENEFICIAL AND PROFITABLE NETWORK UPGRADE PATHS

I B Crabtree and D Munaf

11.1   INTRODUCTION

The basic task is to produce a plan of maximum cost-effectiveness that upgrades switches in a communications network. While upgrading all switches immediately achieves maximum customer satisfaction, it is prohibitively expensive. The task is clearly a constrained resource allocation problem. This chapter describes briefly a number of novel approaches to searching that are suited to this class of optimization problem, and shows the results obtained using simulated annealing on a large multi-user network. The integration of this technique into a tool to support this type of network planning is also described. In terms of problem modelling, as outlined in Chapter 1, the tool performs the search for the optimal solution and also allows sensitivity analysis and what-if analysis to be performed on resulting solutions.

The search techniques discussed in this chapter are all examples of stochastic search techniques — they all incorporate some random element into their search for a solution. This randomness in search, which may appear at first glance to be odd, provides the key to explore effectively very large search spaces that are characteristic of this kind of problem. The next section briefly outlines the features of genetic algorithms, hillclimbing, tabu search, and simulated annealing.

11.2 STOCHASTIC SEARCH TECHNIQUES

11.2.1 Genetic algorithms

A genetic algorithm (GA) is a stochastic and iterative search procedure modelled on the principles of natural evolution. Just as species evolve by 'survival of the fittest', genetic algorithms start with a pool of candidate solutions (the gene pool — a set of genotypes) which 'survive' by being combined with each other (produce offspring) according to some defined operators. Operations on candidate solutions (parents) preserve some characteristics (genetic material) from both parent solutions in each child (inheritance). Candidate solutions are selected for reproduction with a bias towards candidates with a high fitness. The basic cycle of a genetic algorithm is depicted in Fig. 11.1.

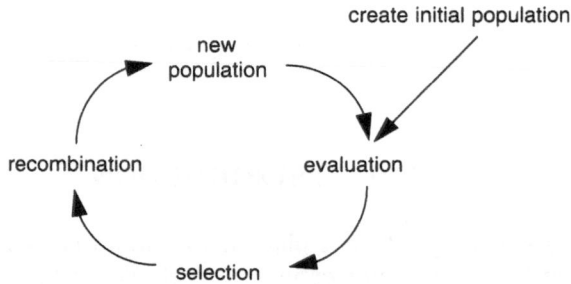

Fig. 11.1   The basic genetic algorithm cycle.

11.2.2 Hill climbing

Hill climbing (HC) starts with an initial random solution and investigates a random sample of neighbouring solutions. The best solution in the neighbourhood is then compared with the current solution, and if the new solution has a lower cost it is accepted as the new current solution, otherwise the current solution is retained and the cycle continues. Since HC only accepts improving moves, it cannot escape from local minima. The HC algorithm is shown in Fig. 11.2.

11.2.3 Tabu search

Tabu search (TS) is a neighbourhood search technique. Given the current solution S and a best solution S^*, a set of neighbouring solutions of S $N(S)$

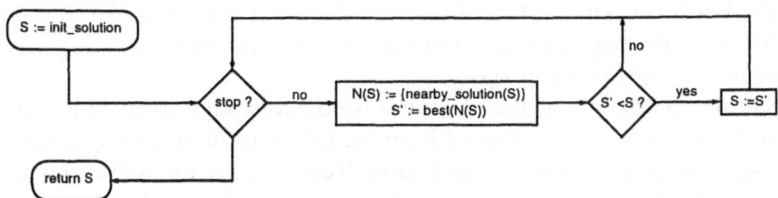

Fig. 11.2   Hill climbing.

is computed, with S' being the best solution in $N(S)$. If S' is better than S, S' is set to S, and if S' is also better than S^*, S' is set to S^*, then the cycle resumes with the new value of S. In this case, S is a local optimum. If S' is worse than S, it may still be accepted as the new S, provided it is not tabu. This allows the search to recover from a local minimum. A solution is tabu as defined by the programmer, for example if the solution has been encountered before in the search process. When the search terminates, S^* is returned, holding the best solution found (see Fig. 11.3).

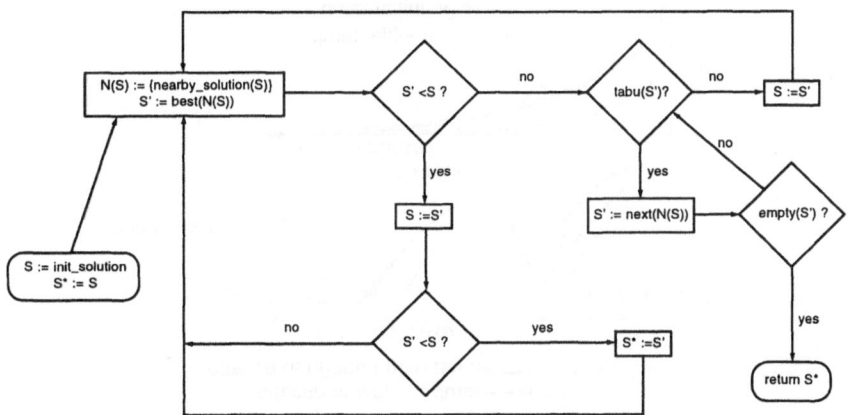

Fig. 11.3   Tabu search.

11.2.4   Simulated annealing

Simulated annealing (SA) is an iterative stochastic technique that finds optimal or near-optimal solutions to complex problems. Here, an optimization problem is seen as analogous to the production of a crystal in physical annealing which starts with a molten solid at a suitably high temperature. The temperature is then slowly lowered. While the temperature is high, the atoms can move around freely and the energy level of the system is high.

But the lower the temperature sinks, and the lower the energy level, the more fixed is the structure of the emerging crystal. Too fast a cooling rate will result in the crystal being flawed.

Simulated annealing, based on the Metropolis algorithm [1], works in the following way — starting with an initial solution and temperature, the annealer executes a random operation from the set of operations on the current solution to produce a new solution and evaluates the change in cost between the previous and the new solution. The higher the current temperature, the higher the chances that a 'backward' step will be accepted, thus providing a way to escape from local minima. The annealer proceeds through many cycles, dropping the temperature at intervals, therefore reducing the probability of accepting backward steps. It is thus important that the annealer is given enough time to converge on a near-optimal solution; if the initial temperature is set too high, the annealer will take longer than is strictly necessary; if it is too low, the annealer may get trapped in a local minimum due to backward steps not being accepted often enough. This process is displayed in Fig. 11.4 with the annealing cycle given in Fig. 11.5.

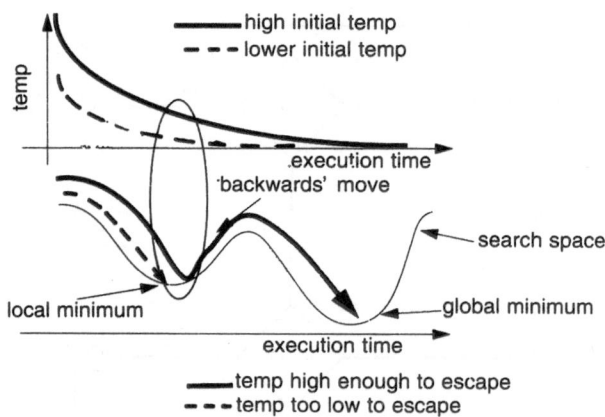

Fig. 11.4   Temperature versus execution time versus solution quality in simulated annealing.

11.2.5   Discussion

A graphical representation of how the four techniques presented here might be expected to perform is given in Fig. 11.6.

Of course, performance of any algorithm is both problem-dependent and implementation-dependent. For example, a simulated annealer's performance will depend on the initial temperature, the cooling rate, and the operators chosen. Nevertheless, assuming suitable implementations of all algorithms,

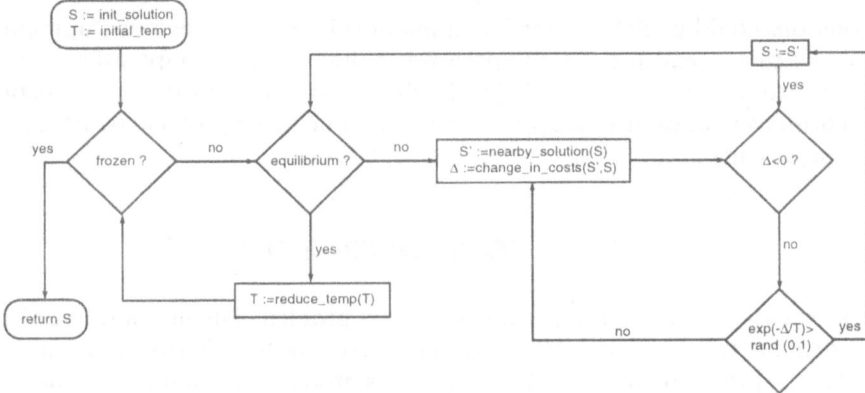

Fig. 11.5   Simulated annealing.

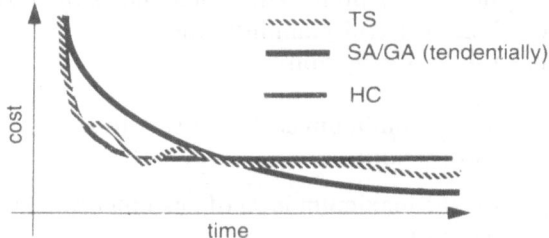

Fig. 11.6   Comparison of different search techniques.

it can be expected that HC will show monotonically decreasing cost of successive solutions, with a more marked decrease at the beginning and a levelling out in later iterations. The final result is likely to be the worst having the highest-cost solution of the four techniques, because it is the only one with no procedure for escaping a local minimum. The other three techniques have such a procedure, although it is different in each. As a result, the cost of successive solutions for them is not monotonically decreasing, and the benefit can be expected to be a better final result than for HC, if given enough time. No attempt has been made to separate simulated annealing and genetic algorithms as far as their performance is concerned, as this is impossible to predict in general. Both would, however, be expected to perform better than TS because their acceptance of worsening solutions is adapted over time, while in TS the strategy for accepting such backward steps remains fixed throughout the search.

For the current problem, both genetic algorithms and simulated annealing would have been a suitable choice, and Buchanan [2] would agree. The problem has a clearly defined cost function, and representation of solutions/chromosomes as strings of switches that can be either on or off (1 or 0) seems natural. Simulated annealing was chosen as operators on solutions could be defined easily, an initial solution was obvious ('upgrade all switches'), and the search space was suitable. In a comparative study produced since (Prosser et al [3]), simulated annealing outperforms genetic algorithms in most instances of a similar combinatorial problem (workforce management).

11.3   TOOL DESCRIPTION

The simulated annealer performs the actual problem solving, but, in order to help the roll-out planning process more effectively, it is advantageous to drive the program and to look at the results produced from different angles reflecting either the customer's or the planner's view. For such situations a graphical user interface is preferred, which allows the user the flexibility they can expect from a useful planning tool.

The tool has the following abilities:

- it can determine the minimum cost necessary to achieve a specified level of customer satisfaction;

- it can determine the maximum level of customer satisfaction achievable with a specified cost;

- it can determine both cost and customer satisfaction levels for any externally proposed solution;

- it can display the impact of an upgrade plan, focusing on any number of customers, any number of switches, the switch location and/or customer sites, or a mixture of these;

- it allows planning in stages, by providing for program states to be saved and restored.

Figure 11.7 gives an indication of the user interface — showing (a) the initial display where the user can select various optimization parameters and change user preferences, and (b) a typical view of an upgraded network for one user. Each box depicts an upgraded switch for that user.

11.4   RESULTS

Experiments have been performed with the tool using historical data representing a network that supports a large number of diverse users. The experiments have focused on ensuring that network upgrades are fair to each

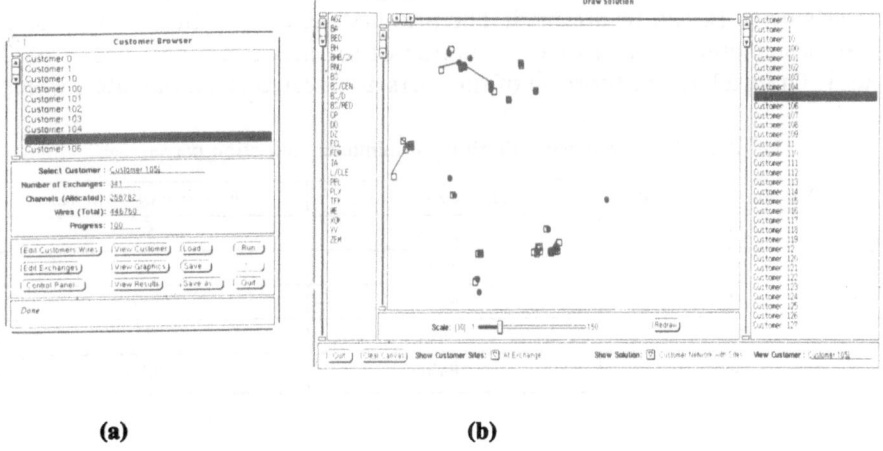

(a)   (b)

Fig. 11.7   Screenshots of the user interface.

user by placing a lower limit on the proportion of the network which must be upgraded to support each of them. The experiments have taken the simplifying assumption that any subset of a user's network may be upgraded. In practice there may be preference constraints that need to be taken into account for each user. These can be taken into account in the tool by specifying certain end-to-end links that must (or must not) be present in an upgrade.

The results are given for a collection of 200 users that together use about 6000 switches.

11.4.1   Setting minimum allocation percentages

Figure 11.8 and Table 11.1 show the results when setting different default allocation percentages. The 'minimum coverage' column in Table 11.1 gives the lower limit on the percentage of each user's network that must be upgraded; the 'achieved coverage' gives the average cover per user; the final column shows the (minimum) number of switches needed to achieve that level of cover.

The dashed line in the graph in Fig. 11.8 indicates the shape the graph would take if the number of switches and channels rose proportionately with the minimum allocation level. It is clear that it does not do so, and upgrading even modest numbers of switches can result in remarkably good coverage for each of the 200 users. With only 187 switches, which represents less than

2% of the whole of the network, an average achieved cover of 45% of each user's network can be attained. The primary reason for these good results can be explained by a close examination of the network topologies of each user. Quite a large proportion of the users are located at similar sites, which

Table 11.1   Coverage with different minimum allocation percentages.

Minimum coverage	Achieved coverage	Number of switches
5%	27%	54
10%	34%	91
20%	45%	187
40%	59%	438
80%	84%	1886

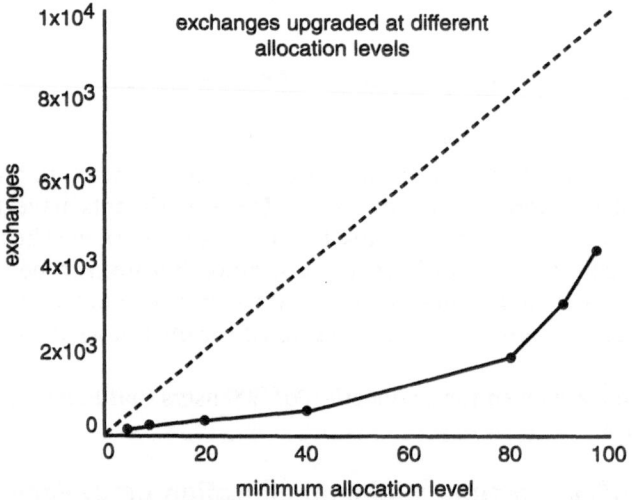

Fig. 11.8   Results with minimum allocation percentages.

tend to be served by the same switches. Choosing these switches ensures that at least one end of the circuit is covered, and, as users tend to share many sites once the 20-30 switch level is exceeded, it is found that a good proportion of the switches are covering both ends of a user's route.

11.4.2   Phased versus non-phased roll-out

Financial considerations require that a major upgrade programme will have to be staged. It is important to select at each stage in the upgrade the most appropriate set of switches to balance the trade-off between fairness to each of the users, and utilization of the new switch. This is the main function of the tool. However, a phased upgrade programme introduces a number of other issues, for example:

- the extent of the initial upgrade;
- the number of incremental upgrades.

Obviously the extent of the initial upgrade and subsequent increments will be determined by current economic conditions, but it is shown here that there are points of maximum benefit if the choice is not too restricted.

It has been seen from the earlier results that upgrading even a small percentage of the switches can lead to quite extensive coverage of each of the users' networks. A measure of the additional benefit of an upgrade can be determined by the total number of effective channels provided over and above what would be expected from a linear extrapolation. Figure 11.9 shows the percentage improvement over linear extrapolation for different minimum allocation levels, derived from Fig. 11.8. It can be seen that the improvement in provision peaks at around 20% minimum provision, which suggests that an initial minimum upgrade of 20% per user would realize the maximum benefit to each user.

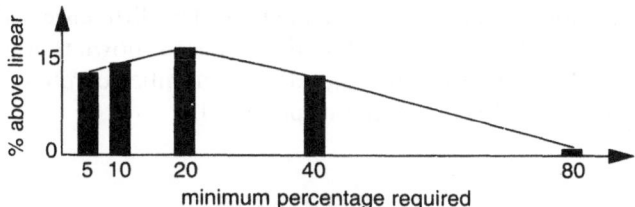

Fig. 11.9   Additional provision as a function of minimum provision requirements.

A phased upgrade, say upgrading $X\%$ of each user's network initially, and then an incremental upgrade of $Y\%$ limits the possibility for a global optimum because the initial upgrade fixes a subset of the switches, and so there are fewer switches to be selected for the incremental upgrade. Upgrading in one step to $(X + Y)\%$ would allow any subset of switches to be selected and this should give better coverage per user. The important issue is the magnitude of this difference. If it took a large number of extra switches to achieve $(X + Y)\%$ coverage by a phased upgrade then it may be advisable to consider a larger upgrade initially. Table 11.2 compares results for planning in two phases against a single phase upgrade.

It is evident that planning in stages results in an overall worse result than can be obtained if upgrading proceeded in one stage. The difference is most marked at very low upgrade percentages and becomes less significant as the upgrade steps are increased. This suggests that the initial upgrade should be larger to make best use of the remaining collection of switches in later phases.

Table 11.2   Results for phased roll-out with minimum allocation percentages.

Minimum coverage (X)	Number of switches (non-phased)	Number of switches (phased from X/2)
10%	91	100
20%	187	194
40%	438	445
80%	1886	1887

11.5   CONCLUSIONS

The tool presented here is a flexible planner for network upgrades. It produces better results than can be obtained through the use of heuristics and has scope for exploring the trade-offs between alternative planning requirements.

It has been shown that there are regions where maximum benefit can be obtained by upgrading a certain number of switches, and also suggested that this may be a good point for an initial upgrade. The difference between phased upgrades has also been quantified, and it has been shown that in many cases it is not significantly worse to upgrade in a number of phases, since each switch can be used to its maximum potential.

REFERENCES

1.   Metropolis N et al: 'Equation of state calculations by fast computing machines', Journal of Chemical Physics, 21 , pp 1087-1092 (1953).

2.   Buchanan J T (Ed): 'Some novel approaches to resource allocation problems', University of Strathclyde, Centre for Network and Resource Management (October 1993).

3.   Prosser P, Muller C and Brind C: 'A preliminary study of stochastic search techniques applied to vehicle routeing problems', University of Strathclyde (February 1994).

12

AN INTRODUCTION TO GROUP DECISION MAKING AND GROUP DECISION-SUPPORT SYSTEMS

S O'Donnell

12.1 INTRODUCTION

Literature on group decision making (GDM) makes frequent reference to the growing need for systems which support groups of people making decisions as opposed to systems which support individuals in making decisions alone — 'the lonely decision maker striding down the hall ... to make a decision is true only for minor decisions. In most organizations ... most major decisions are made collectively' [1]. This is certainly borne out by a survey of decision making in BT carried out in April 1992, which revealed that approximately 70% of respondents feel they make decisions in groups rather than individually [2]. The survey also highlighted that certain types of decisions were usually addressed in groups (e.g. project management, marketing, software development), whilst others were usually addressed alone (e.g. personnel, finance).

Given that most decision-support systems (DSSs) are still aimed at individuals (even though some may be used by groups of people), a review of literature on GDM and group decision-support systems (GDSSs) was undertaken. This chapter introduces material from this review, giving readers an initial insight into the field. From the review, it is notable that very few

GDSSs are in use in commercial settings; while it is likely that a number of factors contribute to this, the fact cannot be escaped that the significant potential claimed for GDSSs has not been realized.

Many of the reported GDSSs support the traditional situation of group members being co-located. However, use of telecommunications facilities provides the opportunity for group members to be geographically separate from one another. The benefits of GDM can then be supplemented by the benefits of distributed meetings (e.g. reduced costs and time for travelling, greater convenience, etc).

GDM and GDSS come under the umbrella term of 'computer supported co-operative work' (CSCW) — the study and theory of how people work together, and how computers and related technologies affect group behaviour. GDSS are usually regarded as being a 'groupware' product, where groupware is the software to support group working [3].

There are a number of differences between making decisions alone and in a group context; these are described in section 12.2 of this chapter. Section 12.3 introduces GDSSs, starting with a definition and then covering the components of such systems, ways of interacting with GDSSs, and how they can help to overcome the disadvantages associated with making decisions in a group. A generic way of comparing the facilities offered by GDSSs is presented in section 12.4, and section 12.5 briefly introduces a number of GDSSs which have particularly interesting features. After presenting many of the potential benefits of GDSSs, section 12.6 looks at practical reasons why these may not have been realized, and provides a number of points of information for anyone intending to develop a GDSS so that the likelihood of the system being successful is maximized.

12.2   GROUP VERSUS INDIVIDUAL DECISION MAKING

12.2.1   Introduction

A reference to 'group' decision making typically brings to mind a picture of group members seated around a table discussing an issue. While this is one type of 'group', other types of group exist, and these are outlined below together with an indication of when each type of group has been found to be most appropriate.

When making decisions in a group setting rather than individually, there are a number of advantages, referred to as 'process gains', some disadvantages, referred to as 'process losses', and some aspects that are simply different. These are discussed below. Some process gains, losses and differences arise specifically when group members use computer-mediated communications (e.g. when group members are physically separate from one another and telecommunications facilities are used) — these are noted.

12.2.2   Types of groups

Management literature refers to three widely discussed decision-making situations.

- Interacting (discussion) groups

    This is the most widely used approach for making decisions in organizations. The typical format for these groups begins with a statement of the problem by the group leader, followed by a somewhat unstructured group discussion for generating information and ideas. The meeting concludes with a majority vote on a consensus decision or opinion.

- Nominal groups

    This is a group meeting in which a structured format is used for decision making. Firstly, individual members silently and independently write down their ideas on a problem or task; then each group member presents one of their ideas to the group without discussion. The ideas are summarized and written up for all to see. After everyone has presented their ideas, the recorded ideas are discussed for the purpose of clarification and evaluation. The meeting concludes with a silent, independent vote on priorities through a rank ordering or rating procedure, depending on the group's decision rule. The group decision is the pooled outcome of individual votes.

- Delphi technique

    Unlike the interacting or nominal group process in which close proximity of group members is assumed (though not essential), participants in the Delphi technique are physically dispersed and do not meet face-to-face for group decision making. The Delphi technique provides for systematic solicitation and collection of judgements of a particular topic through a set of carefully designed sequential questionnaires, interspersed with summarized information and feedback of opinions derived from earlier responses.

A comparison of the relative effectiveness of interacting versus nominal group processes indicated that the optimal combination for a problem-solving committee is:

- to use a nominal group process for fact finding and idea generation;
- to use an interacting group process for information synthesis, idea evaluation, and group consensus;
- to use a nominal group process for the final decision.

12.2.3   Process gains

There are several advantages associated with making decisions in groups, and these are given below.

- Greater number of alternatives

  Kowitz [4] notes that 'in general ... groups will generate more alternatives to a task than individuals working separately'. However, this finding is controversial; other studies indicate that more ideas are generated by individuals working alone than within a group.

- More detailed level of analysis

  People as a group are more likely to identify the hidden strengths and weaknesses of alternatives proposed than individuals working separately.

- Error correction

  The interaction of members within a group provides an error-correcting process, so that flawed solutions can be identified and discarded — often at an earlier stage than for members working individually.

- Identification of 'correct' solution

  Research findings indicate that group performance is often better than that of the average group member in making the correct or optimal decision. One possible reason for the greater effectiveness of group decision making could be that a group is more likely to contain at least one member who can identify the correct course of action. However, Thierauf [5] refers to work which shows this is not the reason for superior group performances — rather it is the interactive process within the group which is critical. Interacting with others in solving a problem appears to contribute to some group effect which influences individual members to approach a problem in ways different from those used when working alone. Additionally, it is becoming increasingly unlikely that any one person will have all the required knowledge and experience necessary to identify the correct solution.

The following process gains arise specifically when people use computer-mediated communications.

• Less socializing and greater attention to task

When people communicate electronically, less socializing takes place, resulting in an increased focus on the task. There is some evidence to suggest that such an increased focus leads to a better quality decision. However, the comments relating to primary tension (see section 12.2.5) should be borne in mind.

• More equal participation

By comparison to face-to-face groups, participation within a group using computer-mediated communications is more equal, participants show less inhibition, and reach decisions which deviate further from initial individual preferences. This increased level of participation may occur principally when anonymous input methods are used, as this has been found to result in greater equality of members of group discussions. In turn, greater participation by all group members reduces the probability of any one member dominating the meeting thereby making better use of all group members.

• Critical evaluation of ideas

People are more critical of each other's ideas when they communicate electronically rather than face-to-face.

Figure 12.1 summarizes these process gains.

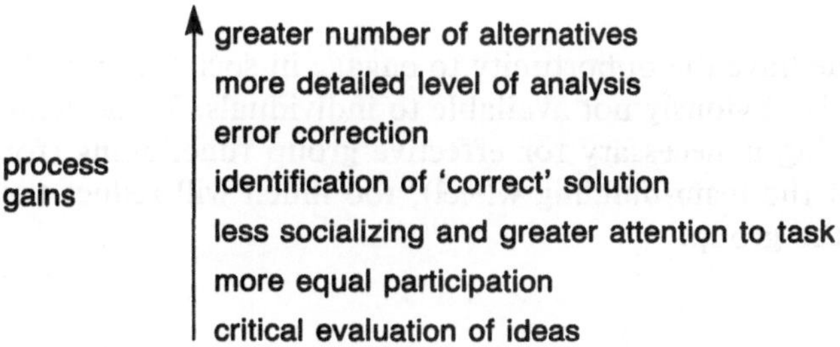

process gains

greater number of alternatives
more detailed level of analysis
error correction
identification of 'correct' solution
less socializing and greater attention to task
more equal participation
critical evaluation of ideas

Fig. 12.1   Summary of process gains.

12.2.4   Process losses

There are several disadvantages arising from making decisions in groups rather than individually, and these are detailed below.

- Production blocking

  This generally refers to the fact that only one member of a group can speak at a time during verbal communication. A number of consequences of this are noted by Valacich et al [6]:

  — attenuation blocking, where group members are prevented from verbalising their ideas as they occur, and consequently they may forget or suppress them because they seem less relevant or less original at a later time;

  — concentration blocking, where group members focus on remembering their own idea instead of generating new ideas based on what is being said by other group members;

  — attention blocking, where group members focus on listening to other members speaking and not on generating new ideas;

  — unequal air time, which refers to the division of available verbal communication time among the group; as groups get larger, the time available to each participant reduces.

- Free-riding

  This refers to the tendency of some group members to rely on the other members to accomplish the task without their contributions. This may be caused by the tendency for people to exert less effort when they pool their efforts toward a common goal than when they are individually accountable, or it may be the result of lack of 'air time' (see above) in larger groups. The concept is discussed further by Valacich [6].

- Socializing

  A group of people have the opportunity to engage in social, non-task, behaviour which is obviously not available to individuals. While some degree of socialising is necessary for effective group functioning (for 'forming' part of the team-building wheel), too much will reduce the effectiveness of the group.

- Domination

  This is where some group members influence others or monopolize the group's time in a manner which is not efficient or effective.

- Groupthink

  Groupthink is a pressure for group members to conform to the consistent thinking patterns of a highly cohesive group. The most often quoted example is the 'Bay of Pigs' invasion, where the USA decided to invade Cuba and were subsequently defeated — the decision was widely held to have been incorrect and to have arisen as a result of groupthink.

- Evaluation apprehension

  This refers to the fear of negative evaluation of ideas shared with the group which may cause individuals to withhold ideas and comments.

- Satisfaction

  This process loss arises specifically when people use computer-mediated communications. In general, these groups express less satisfaction with the group process than groups meeting face-to-face.

Figure 12.2 summarizes these process losses.

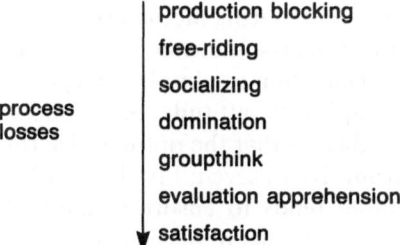

process
losses

production blocking
free-riding
socializing
domination
groupthink
evaluation apprehension
satisfaction

Fig. 12.2   Summary of process losses.

12.2.5   Process differences

Process differences arise simply from making a decision in a group rather than individually.

- Group polarization

  A commonly-held belief is that decisions made within groups tend to converge to the 'average' position of the members of that group. Early work in this field by Tajfel [7] indicated that a group would actually tend to reach a 'riskier' decision than if individuals made the decision alone. Furthermore, once away from the group situation individuals still tended to make riskier decisions than their original ones although not as risky as within the group. Further work has shown that decisions made within a group tend to be more extreme versions of the average views of the individuals within the group. For example, if a group of individuals feel, on average, slightly in favour of a given course of action, then coming together in a group to discuss the issue will result in a more extreme form of that course of action.

● Group size

Thierauf [5] notes that as group size increases, the number of potential information exchanges rises geometrically and the frequency, duration, and intimacy of the information exchange all decline. Kowitz [4] refers to work showing that fewer people interact in larger group discussions; furthermore, aggressive people tend to dominate the conversation and stifle the comments of the other group members. Typically, consensus becomes harder to achieve and affectional ties and satisfaction with the group decline. There is a greater interest in giving information and suggestions and less interest in asking opinions, giving opinions, or showing agreement. Smaller groups tend to resolve differences whereas larger groups tend to use humour as a tension-reducing mechanism. These differences imply that GDSS design must vary somewhat across small and large groups. While no single recommendation can be given on group size because so much depends on the group's task, attitudes of members, etc, Kowitz [4] refers to work which indicates that the optimal size for most problem-solving groups ranges from five to seven. Five is regarded as particularly good in that this number tends to ensure a range of opinions and ideas while still allowing each member to participate freely. The use of an odd number of group members also prevents an impasse situation occurring, where different options are supported by the same number of group members.

● Group history

Group history refers to the length of time the members have been meeting as a group. The distinction between groups with a history and without a history has an implication for the manner in which members interact. Kowitz [4] refers to Bormann's description of primary tension. Primary tension refers to anxieties and worries associated with getting acquainted with new people. Groups without a history tend to experience greater degrees of primary tension because group members lack the necessary information to predict how the other group members will behave. Consequently, interaction patterns in groups without a history during the early stages of discussion are characterized by tentative comments, long pauses, and short messages. Gradually, as the group members get to know one another, the primary tension decreases to a point where the group can work productively. Groups without a history should spend the initial phases of their discussion engaging in communication designed to reduce primary tension. Even groups with a history experience primary tension for a while each time they meet. The initial stages of the discussion should be directed to releasing primary tension. However, care should be taken to avoid an extended period of releasing primary tension as members will experience a frustration because the group is not making progress towards solving its problem.

- Time required

  Group decisions can require more time to make than those made by average individuals taking the same decision alone. For this reason, they can be more costly to make if the process is not adequately controlled.

The following process differences arise specifically when people use computer-mediated communications.

- Member proximity

  The use of computer-mediated communication allows for group members to be physically further apart from one another. Many studies have shown a negative relationship between physical proximity (both actual and perceived) and group cohesiveness. This variable becomes particularly important if strong morale and co-operation are important to the organisation. The use of face-to-face meetings interspersed with those members who are physically separate from one another would help to overcome any such negative effects. This 'member proximity' factor may have implications where people are working alone for extended periods — for example, where people 'telework', group cohesiveness would be increased by the use of reasonably frequent face-to-face meetings.

- Flaming

  There tends to be increased levels of conflict between group members where computer-mediated communication is used. Members tend to become more blunt and assertive in their comments. They tend to express themselves more forcefully and are often not as polite when interfacing in this way compared to face-to-face meetings. This type of behaviour is referred to as 'flaming'. Sproull [8] notes that this situation probably occurs because reminders of the group setting and the other group members are weak in comparison to face-to-face groups, leading to members feeling less bound by conventions and norms appropriate to group meetings. It may be expected that flaming would be less evident in decision groups using videoconferencing, although there are still fewer group cues in this setting than in face-to-face meetings.

Figure 12.3 summarizes these process differences.

process differences

Group polarization	Group size	Group history	Time required	Member proximity	Flaming

Fig. 12.3   Summary of process differences.

12.3   GROUP DECISION-SUPPORT SYSTEMS

12.3.1   Definitions and aims

A variety of definitions and aims of a GDSS appear throughout the literature. The following conveys the essential aspects of such systems:

> '...GDSS aims to improve group decision making by removing common communication barriers, providing techniques for structuring decision analysis, and systematically directing the pattern, timing, and content of discussion to produce higher quality decisions' [5].

12.3.2   The components of a GDSS

A GDSS consists of a number of components:

- hardware;
- software;
- people;
- procedures;
- telecommunications equipment and facilities for distributed group decision making.

Hardware

Typically, each participant in the group has their own input/output device and processor, and an individual monitor. All participants can see a common viewing screen (known as the 'public screen'). The most preferable hardware set-up allows each participant to work independently of the others (through their own monitor), and to publicly demonstrate personal work on the public screen when required. The public screen is typically a display at the front of the room that is seen by everyone in the conference. Its purpose is to provide a common focus around which the discussion flows. The content of the screen is determined by the method of interacting with the GDSS (see section 12.3.3). Typically, a single PC screen is not sufficient to display the required amount of information. This is especially true where windowing is required to compare several alternatives simultaneously. The Plexsys system (see section 12.5.2) uses multiple public screens, which allows one or more screens to display reference information (e.g. historical performance information) while another has currently changing information (e.g. current discussion on a topic); another use is for one screen to show the base case while others show various 'what if' scenarios. The location of the public screen is also a matter for consideration. Usually, the screen is at the front of the room, with participants arranged in one or more rows facing the screen. While this is reasonable for a large group of people, this seating arrangement reduces eye-to-eye contact. For smaller groups, a U-shaped seating arrangement can be used; however, this means that those participants near the screen are also at an angle to it — not an ideal situation. The use of a set of public display screens may help to overcome any of these problems, although this obviously increases the cost.

Software

The software must support both the individual and the group, the process of group decision making, and any specific tasks. Individuals should be supplied with the usual collection of tools (e.g. file creation, spreadsheets, etc). Group support should include:

- summarization of group members' ideas and votes;

- programs for calculating the weights for decision alternatives;

- anonymous input and recording of ideas;

- text and data transmission between group members (including the facilitator), and between group members and a central computer processor.

People

This covers both the participants and the group facilitator (where one is used). Further information on the role of the facilitator can be found in section 12.3.3.

Procedures

These cover the operation of the GDSS, and the effective use of the technology by the group members. Procedures may also cover rules regarding verbal discussion among group members, and the sequence of events during a group meeting.

Telecommunications equipment and facilities

Various equipment and facilities could be used to enable group members to be geographically separate from each other (e.g. ISDN lines, videoconferencing, electronic mail, etc). These would be used to support the software, the people within the group, and the procedures used.

A common arrangement of these components is the decision room — the electronic equivalent of a traditional face-to-face meeting. A conference room has terminals for each participant, providing various levels of support (the decision-making process may take place verbally or use electronic communication between group members). Group members communicate directly with other group members and/or the public screen. The screen can be used to display ideas, show analyses of group preferences and votes, and display decision aid tools (spreadsheets, graphs, etc). It is the most common type of GDSS. The Colab room at Xerox Palo Alto Research Center is a much quoted example of a decision room (see section 12.5.1).

A typical arrangement of equipment for use in a decision-room GDSS is shown in Fig. 12.4.

12.3.3   Interacting with GDSSs

There are a number of processes by which group members can interact with a GDSS — using a chauffeur, through a facilitator, directly (i.e. a user-driven system). Each of these are briefly described below.

- Chauffeur-driven interaction

    In this approach, one person in the meeting takes orders from the group members and implements them using the GDSS technology. The chauffeur may, but need not, be a group member. The chauffeur may be the only person to have access to the technology, or the group members may have only very limited features of the technology available to them.

- Facilitated interaction

    This is the most frequently used method of interacting with a GDSS. A facilitator assists the group in using the GDSS technology. In one form,

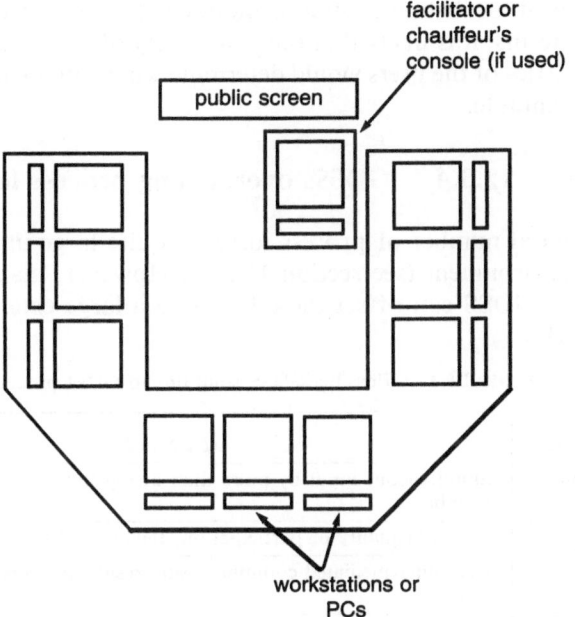

Fig. 12.4   Typical equipment set-up for a decision-room type GDSS.

the facilitator is the only person to have access to a workstation (usually accompanied by a large screen display visible to all group members). The second form allows each member of the group access to a workstation but the facilitator directs the group members as to what system features to use and when to do so. In some cases, the facilitator has system features available that are not available to all members of the group. The difference between a chauffeur and a facilitator is that the former only 'drives' the system according to the wishes of the group members whereas the latter directs the use of system features.

- User-driven interaction

  Each member of the group is equal regarding the access to the GDSS. Each member has a workstation and full system features. Output from an individual's personal screen can be sent to the public screen at any time. It should be noted that users must be trained to use this type of system.

In a comparison of the three methods of interacting with a GDSS, Dickson et al [9] report that chauffeured and facilitated groups achieved higher degrees of consensus than user-driven groups (on a task involving the allocation of a substantial sum of money between six worthy projects). In practical terms, it is likely that the complexity of the decision together with characteristics of the users would determine which interaction method would be most suitable.

12.3.4   GDSSs overcoming process losses

The potential number of process losses is quite large in a group decision-making environment (see section 12.2.4). However, many of the facilities offered by GDSS can offset these losses to a large extent, as indicated in Table 12.1.

Table 12.1   Possible GDSS solutions to GDM process losses.

Process loss	GDSS solution
Production blocking	Simultaneous input by more than one group member
Free-riding	Greater equality of participation, facilitated by anonymous input
Socialising	Computer-mediated communication results in a greater focus on the task
Domination	Equality of participation, facilitated by anonymous input
Groupthink	Non-conformity, stimulated by anonymous input
Evaluation apprehension	Reduction of apprehension, facilitated by anonymous input

12.3.5   Analysis

The use of a GDSS to support group decision making can be justified if it can provide one or more of the following:

- increased efficiency of meetings;

- improved quality of meetings and/or decisions;

- improved processes.

Increased efficiency could be achieved by reducing the total amount of time needed for a meeting, where time required is calculated as number of members multiplied by the duration of their input. It is not obvious that this gain has been demonstrated for GDSS.

The 'quality' of a meeting or a decision is difficult to define. Typically, it is assessed by the level of participation of all group members. The quality of a decision is also assessed by looking at the number of alternatives proposed, the quality of these alternatives (again, this is difficult to assess), the level of commitment to the decision, and/or the quality of the decision as judged by the group members. On the majority of these dimensions, a GDSS does appear to improve the quality of decisions.

Improving the process of running a meeting is another potential benefit that is difficult to quantify and measure. However, as an example, if two people can contribute to a meeting at the same time (i.e. in parallel instead of sequentially), then the process of the meeting could be said to be improved.

The use of telecommunications facilities allows for group members to be physically separate from each other, introducing the possibility for considerable savings on travelling (time, cost, etc). This justification for GDSSs is valid now; however, improvements in facilities are required and these would allow this justification to be more fully realized. It may also contribute to the increased efficiency of meetings.

GDSS appears to offer the most potential for improving the quality of a decision, improving the group decision-making process and significantly reducing travelling time.

12.4   COMPARING FACILITIES OFFERED BY GDSSs

There are a large number of reported GDSSs, and it would be useful to be able to compare the facilities offered by them. Although a number of frameworks for this appear in the literature, these were found to be of limited use in that they either focused on the physical setting for the GDSS, or else the 'granularity' of comparison levels was inappropriately large.

A framework used by the author for comparing the support offered by GDSSs is presented below. It is based on the problem-solving process (a logical sequence of steps for solving problems and improving the quality of decisions [10]) and indicates the extent to which GDSSs support steps in this process.

- Identify the problem

  This is widely regarded as one of the most important stages of decision making — achieving a common understanding of the issue is crucial to the group discussing the issue in a constructive manner. A small number of GDSSs support this stage.

- Gather data

  This stage is supported by many, although not all, of the GDSSs. It covers the gathering and displaying of information pertinent to the decision.

There is little in this stage that is specific to group as opposed to individual decision making — displaying the information is the main difference and this is covered as part of the discussion on what is displayed on the public and private screens.

- Analyse data

  Again, many GDSSs offer facilities for this although they are the same as those offered to individuals carrying out such analysis alone. Facilities typically cover spreadsheet packages, modelling software and 'what if' calculations.

- Generate solutions

  Typically, this may occur using the brainstorming technique. However, the value of using or developing a GDSS facility to support the generation of solutions is questionable. Pollock and Kanachowski [11] refer to work showing that pooling suggestions made by individuals produced more suggestions than if the same number of people worked as a group in a brainstorming session.

- Analyse the alternative solutions

  Analysis is widely regarded as one of the most productive stages of decision making that can be carried out within a group. Without some form of computer-based support, it is often reported as being unstructured and less productive than it could be. Given the scope for computer-based support of this stage, the facilities offered by GDSSs appear sparse. A GDSS which provides a straightforward, simple way of supporting a group of people analysing various alternatives was not identified in the literature reviewed.

- Select the preferred solution

  Many of the GDSS tools support this aspect of decision making. Examples cover simple support for voting-type behaviour [12], ranking behaviour [13], and decision matrices where the weighting attached to particular criteria can be varied to examine the impact on various alternative solutions [14].

Another useful dimension for comparing GDSS would be based on the types of decisions they could support. As indicated in the introduction, people typically make certain types of decisions in groups, and it would be useful to know if certain GDSSs were developed to support specific types of decision. Unfortunately, no reference to this was found in the literature surveyed — indeed, no reference was found to the possibility that certain types of decisions are more appropriately made in groups than others.

12.5   A REVIEW OF SELECTED GDSSs

As indicated above, there are a large number of reported GDSSs, many developed from and used within academic settings, and a few from commercial companies. A small number of particular interest are briefly described below.

The Lotus Notes product is currently receiving much interest, and could conceivably be used as a group decision making aid. However, it is not obvious that its strength lies in such support — it is primarily a groupware product and offers group decision facilities that could be expected from groupware/electronic mail products (for a description of Lotus Notes, see Golfin and Jackson [15]).

12.5.1   The Colab system

The Colab system [16, 17] has been developed by Xerox at the Palo Alto Research Center. It is widely regarded as a sophisticated system, offering a wide variety of features.

The major feature is the facility to structure an electronic discussion (the Cognotor) into brainstorming, organizing and evaluation phases so that all the ideas contributed by participants can be recorded, grouped into appropriate subject categories, and assigned relative priorities. This enables group members to outline the task confronting them so that it can be systematically addressed.

The tool also offers:

- the Boardnoter (an electronic chalkboard);

- the Argnoter, a kind of argumentation spreadsheet that displays proposals for action from group members and permits them to be evaluated in relation to the points of view represented in the group.

Progress towards a group consensus is displayed on the public screen.

12.5.2   The Plexsys system

The Plexsys system [18, 19] has been developed at the University of Arizona Planning Laboratory using equipment manufactured by NCR. The software is available for purchase, and has been transferred to university and corporate settings. The Plexsys system is of particular interest because it provides an extended data gathering facility, allowing external data to be integrated into key items. It also provides the following tools:

- a session director, which acts as guide to tool selection for the facilitator, and sets the agenda;

- electronic brainstormer, which supports simultaneous and anonymous idea generation;

- issue analyser, which aids in identifying key items from idea generation and integrates external information relating to key items;

- voting;

- topic commenter, which supports idea solicitation on a list of topics;

- policy formation;

- organizational infrastructure, which provides support for identification of organizational characteristics;

- stakeholder identification and assumption surfacing, which supports the systematic evaluation of implications of proposed policy or plan;

- alternative evaluator, which provides multicriteria decision-making support.

12.5.3   The SAMM system

The Software Aided Meeting Management (SAMM) system [9] has been developed by the University of Minnesota. Its particular strength is the support provided for problem definition, i.e. arriving at an agreed definition of the problem. It also provides the following tools:

- establishment of decision criteria;

- definition of alternatives;

- evaluation of the alternatives (rating, ranking and voting

12.5.4   The Claremont system

The Claremont system [19, 20] has been developed by Claremont Graduate School (USA). It uses equipment manufactured by HP. It is interesting in that the system was designed to investigate the advantages and disadvantages of using a touchscreen as an interface — the design objective was to create as nearly 'typewriterless' a system as possible for the meeting participants. Experience with a number of groups and individuals who have used the voting facility of the system shows that the touchscreen requires almost no learning time.

It differs from many other systems in that the majority of typing is done by a chauffeur, who guides the meeting through 'choice-making' software.

The Claremont system provides the following tools:

● information sharing;

● creating ideas;

● making choices (this is the major component of the system.

12.5.5   The VisionQuest system

The VisionQuest system [21] has been developed by OmniQuest Software Inc, Austin, Texas.

VisionQuest is designed to facilitate collecting, analysing, communicating and synthesizing the viewpoints and judgements of people working together in teams. It provides the following facilities:

● meeting agenda and structure setting;

● electronic nested and hierarchical idea generation (an updated version of brainstorming);

● setting priorities through ranking and voting;

● monitoring the overall effectiveness of meeting presentations;

● information integration (e.g. the easy movement of data between or among the software components) and information storage, display and reporting (e.g. the generation of summary reports and meeting minutes).

A recent review of this sytem [22] highlighted difficulties in importing material from other applications such as spreadsheets or scanned documents, and concluded 'even if you're facing disaster in the conference room ... this isn't your salvation'.

12.5.6   The TeamFocus system

The TeamFocus provides the following facilities:

- electronic brainstorming (i.e. idea generation);
- issue identification, review, editing and categorization;
- issue consideration and ranking;
- voting on and ranking the importance of issues using comments to explore issues in an organised manner;
- policy formation and development.

12.5.7   The GroupSystems V system

The GroupSystems V system [23] has been developed by Ventanna Corp, Tuscan, Arizona.

GroupSystems V supports group decision making across dispersed geographic locations. Users can converse privately, on conference calls, or broadcast a message to all team members independent of the GroupSystems program. Distributed meeting capabilities include the building of an information base with idea generation and organization, alternatives evaluation and consensus building, and decision making. An entire record of the meeting is provided to each participant. Hewlett Packard have been using this system within an electronic meeting room in Palo Alto, California since March 1992, and report [24] that it has cut meeting times by up to 60% by helping to increase consensus and reduce company politics. It is planning to expand its usage to other sites throughout the world.

12.5.8   The MacPolicy system

The MacPolicy system [25] has been developed in the future and policy planning unit of the Free University of Brussels. The system is interesting in that it is based on the Delphi technique (see section 12.2.2). It supports groups of people who are geographically distant from each other and also removes the need for people to meet together at one time.

The MacPolicy system provides the following facilities:

- support for structured discussion, using an iterative consensus-seeking questioning technique with anonymous information feedback;

- a database-mailing system that enables users to send and store questions, reports, opinions and critiques;

- statistical tools to analyse collected data;

- multicriteria decision aids.

12.6 CONCLUSIONS

Tasks that involve generating ideas or plans, solving problems, and resolving conflicts of different viewpoints or interests are particularly good candidates for effective GDSS utilization. Those that involve resolving conflicts of power or group leadership, and those that are more related to actual detail execution, that more often resides with individuals, are less effectively addressed by a GDSS.

Given the rise in the number of decisions made by groups within organizations, the potential for GDSS is great. However, as yet this potential does not appear to have been realized, and there do not appear to be many commercial off-the-shelf GDSSs. Some of the reasons for this are as follows.

- In practice, very little innovative or inventive work has been done — '... little new has been designed. The applications are still mechanizations of the way groups have always worked. The focus is on text-oriented tasks' [19].

- Support for inappropriate parts of meetings — Gray [19] notes 'the software developed for GDSS focuses almost exclusively on assisting brainstorming and mechanising voting, two of the rarer events in business meetings'.

- Lack of value to organizations — Gray [19] notes that 'the history of GDSS in real organizations is not encouraging. The inability of systems to survive beyond the whims of an individual champion may imply that they do not do anything that is sufficiently important to an organization to maintain its investment'.

- Lack of attention to the way people actually make decisions — Pollock and Kanachowski [11] note that 'the indiscriminate application of normative theories to the design of systems may be inappropriate if people do not behave according to the normative principles ... systems can be improved by integrating descriptive and normative theories'.

Kraemer and Kind [26] note that 'in short, outside of research environments, GDSSs currently remain more prospect, promise and possibility rather than successful operations'. Furthermore, results of a survey reported in the Financial Times in 1993 [27] indicated that 86% of people surveyed did not think the time was right to act on groupware generally. A telling quote is also given in this report from an associate partner at Andersen Consulting: 'the biggest gap is in the lack of a telecoms infrastructure. If you really want to get groups working together, they have to be free to do it from wherever they are. At the moment, there are too many compromises made because of the lack of a telecoms capability.'

Despite these reservations, the drivers for GDSSs are increasing both in number and importance. These drivers include:

- the dispersion of a company's workforce as more companies go global;

- the downsizing of companies;

- the rapid turnover in a company's workforce;

- the need to reduce the amount of travel;

- the sheer increase in pace of today's world;

- the trend for companies to become more transient;

- the rise in the number of inter-company projects, often of relatively short duration;

- no one person can know everything;

- the continuing pressure to arrive at better decisions, faster.

In order to maximize the chances of success for any GDSS, developers should bear in mind the following points:

- the views of the group members — a GDSS is obviously likely to be more successful where the group members are enthusiastic to use the GDSS and are informed of the potential benefits;

- the presence of a 'champion' for the system — there needs to be someone to highlight the benefits of the GDSS and to push through its introduction and set-up;

- the 'look and feel' of the surroundings that the GDSS is to be used in — be this a decision room for co-located group members or a terminal where the group is geographically separated;

- be aware of situations in which GDSS is appropriate — these are when speed and cost are less important than avoidance of mistakes, quality of the decision, and/or group acceptance of the decision;

- be aware of the appropriate type of group for each stage of the problem-solving process — the optimal combination of group processes for a problem-solving committee is:

  — to use a nominal group process for fact finding,

  — to use an interacting group process for information synthesis, idea evaluation, and group consensus;

  — to use a nominal group process for the final decision (see section 12.2.2 for a description of nominal and interacting groups);

  and then ensure that any GDSS software developed is appropriate to the type of group;

- be aware of the effects of increasing group size, e.g. increased information exchanges, with decreases in the frequency, duration and intimacy of information exchanged, and different approaches to resolving differences;

- provide facilities to decrease the primary tension within a group;

- aim to provide the appropriate number of features on a GDSS — there should be sufficient features that the system is worthwhile using, but the more features there are, the more training that is required and, if the system is to be used by the same participants frequently, a reasonable level of training may be acceptable, but, if participants only use the system infrequently, then the system must be easy to use with the minimum amount of training;

- provide facilities for group members to assess their progress towards group goals — for example via the use of, and referral to, an agenda.

In summary, the potential for GDSSs is significant and is likely to increase. However, the development and introduction of such systems needs careful consideration.

REFERENCES

1. Turban E: 'Decision support and expert systems — management support systems', Macmillian Publishing Company (1990).

2. O'Donnell S: 'Survey of DSS opportunities in BT: questionnaire results', IT00522:BT__SURV:DOC:009, Issue 1 (30 October 1992).

3. Rogers A (Ed): 'Groupworking over networks', Special Issue, BT Technol J, 12 , No 3 (July 1994).

4. Kowitz A C and Knutson T J: 'Decision making in small groups', Allyn and Bacon (1980).

5. Theireuf R J: 'Group decision support systems for effective decision making', Quorum Books (1989).

6. Valacich J S, Dennis A R and Nunamaker J F Jr: 'Electronic meeting support: the groupsystems concept', International Journal of Man-Machine Studies, 34 , pp 261-282 (1991).

7. Tajfel H and Fraser C: 'Introducing social psychology', Penguin Books (1981).

8. Sproull L and Kiesler S: 'Connections — new ways of working in the networked organisation', MIT Press (1991).

9. Dickson G W, Lee J E, Robinson L and Heath R: 'Observations on GDSS interaction: chauffeured, facilitated, and user-driven systems', in Blanning R and King D (Eds): 'Proceedings of the Twenty-Second Annual Hawaii International Conference on System Sciences', 3 , pp 337-343 The Computer Society Press (1989).

10. 'Meeting Customer Requirements — a pocket guide to continuous improvement', BT Group Quality Unit (1991).

11. Pollock C and Kanachowski D: 'Application of theories of decision making to group decision support systems', International Journal of Human-Computer Interaction, 5 , No 1, pp 71-94 (1993).

12. Gear A E and Read M J: 'On-line group process support', International Journal of Management Science, 21 , No 3, pp 261-274 (1993).

13. Valachich S J, Vogel D R and Nunamaker J F Jr: 'Integrating information across sessions and between groups in group decision support systems', in Blanning R and King D (Eds): 'Proceedings of the Twenty-Second Annual Hawaii International Conference on System Sciences', 3 , The Computer Society Press (1989).

14. Swap W C (Ed): 'Group decision making', Sage Publications (1984).

15. Golfin N G and Jackson M: 'A groupware trial in BT', BT Technnol J, 12 , No 3, pp 51-55 (1993).

16. Tatar D G, Foster G and Bobrow D G: 'Design for conversation: lessons from Cognotor', International Journal of Man-Machine Studies, 34 , pp 185-209 (1991).

17. Nunamaker J F Fr: 'Group decision support systems (GDSS): past and future', in Shriver B D (Ed): 'Proceedings of the Twenty-Second Annual Hawaii International Conference on System Sciences', IEEE Computer Society Press (1989).

18. Buckley S R and Yen D: 'Group decision support systems: concerns for success', The Information Society, 7, pp 109-123 (1989).

19. Gray P and Olfman L: 'The user interface in group decision support systems', Decision Support Systems, 5, pp 119-137 (1989).

20. Mandviwalla M, Gray P, Olfman L and Satzinger J: 'The Claremont GDSS support environment', in Milutinovic V and Shriver B D (Eds): 'Proceedings of the Twenty-Fourth Annual Hawaii International Conference on System Sciences', IEEE Computer Society Press (1991).

21. Alavi M: 'Group decision support systems — a key to business team productivity', Journal of Information Systems Management, pp 36-41 (Summer 1991).

22. Phillips K: 'Brainstorming on-line — collaborative technologies VisionsQuest for Windows 2.11 workgroup office automation software', Corporate Computing, 2, No 5, pp 50-51 (May 1993).

23. 'Groupsystems V: Ventana Corp's distributed group productivity software', Software Magazine, 12, No 12, p 67 (1992).

24. 'Can groupware put you in the winning team?', FT Business Computing Brief, 237/6-237/7 (December 1993).

25. Kenis D and Verhaegen L: 'The MacPolicy project: developing a group decision support system based on the Delphi method', in Bots P W G, Sol H G and Traunmuller R (Eds): 'Decision Support in Public Adminstration', Elsevier Science Publishers, North Holland (1993).

26. Kraemer K L and Kind J L: 'Computer-based systems for co-operative work and group decision making', ACM Computing Surveys, 20, No 2 (June 1988).

27. Sedacca B: 'An aid to teamwork for PC-users: groupware', Financial Times, p VI (26 October 1993).

17. Marschak, T. for. Hondo decision support systems (GDSS), part one. Acme Shorter, 1992 (math), proceedings of the Twenty Second annual Hawaii international conference on system science, IEEE Computer Society Press, 1990.

18. Marschak et al., interview to group decision support systems, Decision Sciences Institute, 7 no. 115, 1 (1989).

19. Vogel, Nunamaker, Congr., Chapter 1, and Satisfaction: The Consensual GDSS outcomes, proceedings of the Twenty Third Annual International Conference on system science, Hawaii, 1990.

20. Poole, M. Information support systems, Academy Management Review, 1990.

21. Warner, interview to group decision support systems, Interface, September 20 October 1991.

22. Watson, Ho, a distributed group decision support system, Information Systems, 16, no. 2, p.6, 1992.

23. Huber, Group decision making, Communications Association, 1989.

24. Brooks, Vogel, Designing electronic meeting systems, Group support in the Twenty First, proceedings Hawaii International Conference on system science, 1992.

25. Gray and group decision making, Academy Management Review, 20, No. 2, 1995.

26. Nunamaker, electronic meeting systems for group work, Communications, 1991.

Index